MW00843462

RELATIVISM AND REALISM IN SCIENCE

AUSTRALASIAN STUDIES IN HISTORY AND PHILOSOPHY OF SCIENCE

VOLUME 6

RELATIVISM AND REALISM IN SCIENCE

Edited by

ROBERT NOLA
Department of Philosophy, University of Auckland, New Zealand

KLUWER ACADEMIC PUBLISHERS
DORDRECHT / BOSTON / LONDON

Library of Congress Cataloging-in-Publication Data

Relativism and realism in science.

(Australasian studies in history and philosophy of science; v. 6)
Includes index.
1. Science—Philosophy. 2. Science—Social aspects.
3. Relativity. I. Nola, Robert. II. Series.
Q175.R385 1988 501 88—4516
ISBN 90—277—2647—7

Published by Kluwer Academic Publishers,
P.O. Box 17, 3300 AA Dordrecht, The Netherlands

Kluwer Academic Publishers incorporates the publishing programmes
of D. Reidel, Martinus Nijhoff, Dr W. Junk and MTP Press.

Sold and distributed in the U.S.A. and Canada
by Kluwer Academic Publishers,
101 Philip Drive, Norwell, MA 02061, U.S.A.

In all other countries, sold and distributed
by Kluwer Academic Publishers Group,
P.O. Box 322, 3300 AH Dordrecht, The Netherlands

Printed in The Netherlands

TABLE OF CONTENTS

FOREWORD

The institutionalization of History and Philosophy of Science as a distinct field of scholarly endeavour began comparatively early — though not always under that name — in the Australasian region. An initial lecturing appointment was made at the University of Melbourne immediately after the Second World War, in 1946, and other appointments followed as the subject underwent an expansion during the 1950s and 1960s similar to that which took place in other parts of the world. Today there are major Departments at the University of Melbourne, the University of New South Wales and the University of Wollongong, and smaller groups active in many other parts of Australia and in New Zealand.

"Australasian Studies in History and Philosophy of Science" aims to provide a distinctive publication outlet for Australian and New Zealand scholars working in the general area of history, philosophy and social studies of science. Each volume comprises a group of essays on a connected theme, edited by an Australian or a New Zealander with special expertise in that particular area. Papers address general issues, however, rather than local ones; parochial topics are avoided. Furthermore, though in each volume a majority of the contributors is from Australia or New Zealand, contributions from elsewhere are by no means ruled out. Quite the reverse, in fact — they are actively encouraged wherever appropriate to the balance of the volume in question.

R. W. HOME
General Editor
Australasian Studies in History and Philosophy of Science

ACKNOWLEDGEMENTS

I would like to thank Rod Home, the General Editor of this series, for his cheerful encouragement from the time when I first suggested to him that I might undertake the project of collecting some papers on the topic of this volume by philosophers who are currently working, or who have worked, in New Zealand or Australia. I would also like to thank Jack Smart, a member of the Editorial Advisory Board for the series, for his assistance in preparing the volume. Others who made helpful suggestions to me include Jan Crosthwaite, John Clendinnen, Fred Kroon and Denis Robinson.

All of the papers were specially commissioned for this volume though earlier versions of some papers have been presented at conferences. In particular, David Papineau's paper 'Does the Sociology of Science Discredit Science?' was originally delivered to the Israel Colloquium for the History, Philosophy and Sociology of Science. David Papineau would like to thank the organizers of the Colloquium and the Van Leer Jerusalem Foundation for their invitation and I would like to thank them for their permission to publish the paper in the present volume.

ROBERT NOLA

ROBERT NOLA

INTRODUCTION: SOME ISSUES CONCERNING RELATIVISM AND REALISM IN SCIENCE

1. INTRODUCTION

Realism is an interpretation of scientific theories nearly as old as science itself. When Aristotle presented Eudoxus' theory of nested concentric rotating spheres devised to account for the motions of the planets and stars, he assumed that it gave a picture of how the cosmos was actually constructed. His immediate successors were less sanguine about the possibility that any scientific theory, whether of concentric spheres or of combinations of epicycles and deferents, could provide us with such a picture at all. They were early instrumentalists who preferred to regard theories as devices whose task was only, as they put it, "to save the phenomena". Realism and instrumentalism are rival views of science; a third rival is relativism.

While realist and instrumentalist interpretations of scientific theory are of ancient vintage, the relativist interpretation is of much more recent vintage, but with roots in a philosophical relativism even more ancient as a doctrine than either realism or instrumentalism. Protagoras is philosophy's first recorded relativist. However in recording his views, Plato, in the *Theaetatus*, delivered the first drubbing meted out to a relativist, a fate to be repeated down the centuries. Our own century has been no less unkind to philosophical relativists starting with early critics as diverse as Lenin[1] and Husserl[2] and continuing to the present with writers such as Passmore,[3] Davidson,[4] Putnam,[5] Newton-Smith,[6] Burnyeat,[7] Rorty[8] and Siegel,[9] to name just a few. No matter how much philosophers are at odds with one another they seem, with only a few exceptions,[10] to be united in their condemnation of relativism. However relativism exerts a strange fascination and a number of philosophers, as Putnam has noted in his own case,[11] have found it instructive to come to terms with the ways they find it to be self-refuting or incoherent (if they find it so at all).

Realism and instrumentalism were the two main rival interpretations of scientific theory discussed in any detail by philosophers before this century; the relativist interpretation hardly featured in their debates.

Robert Nola (ed.), Relativism and Realism in Science, 1—35.

Long an outcast from philosophy, relativism has, for most of this century, found a home in the doctrines of most (but not all) sociologists of knowledge and, more recently, in the doctrines of most (but, again, not all) sociologists of science, several historians of science and a handful of philosophers of science. Just how much of relativism they have sheltered varies considerably; in some cases most of the main features of relativism have been adopted while in a few cases only the name has been retained, perhaps because of its shock value. Under the influence of many of these theoreticians relativism has become a pervasive interpretation of most aspects of the scientific endeavour. Suffice to mention that relativism has been espoused, in one form or another, at least with respect to (a) the methods and canons of reasoning found in science, (b) the ontology of science (the alleged incommensurability of pairs of theories being one ground for this claim), (c) observations in science (the alleged theory-ladenness of observations and the unavailability of a foundationalist epistemology being two grounds for the claim of observational relativity), (d) truth claims for the theoretical statements of science (one ground for this being the rejection of realist theories of truth such as the correspondence theory of truth).

In general, many sociologists and some historians and philosophers of science have regarded science as a cultural effusion essentially no different from any of the belief systems and practices with which it has been traditionally contrasted, such as religion or myth. They have also alleged that science has no privileged status as a means of gathering knowledge of the world, as many philosophers have claimed, and have insisted that it ought to be studied in the same way as one would study any cultural phenomenon. Just as anthropologists have recorded the beliefs and practices of alien tribes of people and have drawn relativist conclusions with respect to both beliefs and values, so sociologists, along with their fellow-travelling historians and philosophers, have entered the domains of tribes of scientists to study their beliefs and practices and have, similarly, drawn relativist conclusions with respect to their beliefs, norms and goals. As a result of these studies they have tended to cast aspersions on the picture of the scientific endeavour traditionally presented by philosophers with a less anthropologically oriented approach to the study of science. Such studies are, of course, a valuable contribution to our understanding of science. But need we accept the relativist conclusions that the sociologists have drawn, especially in the case of the tribe of scientists? Moreover, what of the status of the lucubrations of the tribes of anthropologists and sociologists themselves? Are the reports of their investigations and the norms governing their procedures merely another cultural effusion like the matters they study, or do they have a privileged status in that they are based on a firmer foundation? These and other questions concerning relativism will be taken up again subsequently.

At first glance it would appear that relativism stands in marked contrast to realism. On the whole this is correct. However if distinctions are carefully drawn between varieties of realism and relativism it will be seen that some of the varieties are compatible with one another. For example, it will be shown below that some sociologists of science are realists in the sense that they maintain that there is one common reality to which scientific theorising is a response (i.e., they are not ontological relativists but ontological realists) while they also maintain that there is no one theory of scientific method or set of canons of reasoning that can justifiably be accepted above any other (i.e., they are epistemological or methodological relativists). Clearly not all brands of non-realism are relativistic in character. In fact given the varieties of realism in science that can be distinguished there emerges a corresponding range of varieties of non-realism of which relativism in science is only one. Thus the realisms and relativisms evaluated here are not exhaustive of the range of possible interpretations of the scientific endeavour but they do, on the whole, stand in contrast with one another.

In this introductory survey of the issues which surround both realism and relativism in science, no attempt will be made to present either a complete taxonomy of the varieties of realism and relativism that can be found or to argue for or against the plausibility or coherence of each kind. What will be offered is a sketch of a number of pigeon-holes into which some of the varieties of realism and relativism can be slotted and, in a few cases, some of the difficulties which face these varieties, most of the difficulties being raised in the case of relativism rather than realism. Some of the claims made by some leading sociologists of science will also be discussed to locate them with respect to the varieties of relativism distinguished. The contributions that the papers collected in this volume make concerning realism and relativism will be summarized briefly in the final section of this introduction.

2. VARIETIES OF REALISM

Realism stands in contrast not only to the already mentioned instrumentalism and relativism but also to phenomenalism, pragmatism, verificationism, reductionism, internal realism, constructive empiricism, and many more besides. Since realism is a fairly protean doctrine, what might seem like an opponent of one kind of realism may become an ally for another kind. A useful grip on the varieties of realism can be obtained by following a broad three-fold distinction between ontological realism, semantic realism and epistemological realism suggested by Geoffrey Hellman and others.[12] It is also possible to establish a similar three-fold distinction with respect to the varieties of relativism; this will be discussed in the next section.

2.1. *Ontological Realisms*

Perhaps the weakest form of ontological realism can be formulated thus:

OR1: There is something which exists in a suitably mind-independent manner.

What is this something? It could range from the undifferentiated Parmenidean goo of being to numbers or Kantian things-in-themselves. None of these seem likely candidates for scientific investigation; so already OR1 needs the qualification that whatever exists is a possible subject for scientific investigation in order to distinguish scientific realism from realisms of a non-scientific sort. However care is needed in formulating this qualification as it may introduce unwanted epistemic notions into a purely ontological formulation of realism (see the following Section 2.3). Because OR1 is the weakest form of realism Devitt has stigmatized it as fig-leaf realism adding that "it is an idle addition to idealism".[13] Idle it may be, but it is a kind of realism with which even Nelson Goodman, as he indicates in a passage entitled 'Relative Reality', might be able to live; he says that it ". . . is a world without kinds or order or motion or rest or pattern — a world not worth fighting for or against".[14] But at least the negation of OR1 yields a significant form of non-realism, viz., subjective idealism. Reformulating OR1 only marginally less weakly in terms of the bare existence of objects yields:

OR2: There are individual objects which exist in a suitably mind-independent manner (and which are open to scientific investigation).

One way of construing the denial of this is, once more, subjective idealism of a Berkeleyan sort. However, an alternative construal would yield a realism committed to, for example, the bare existence of fields (of which objects are some kind of construction); thus not all theses which rival OR2 need be about items that a scientific realist would necessarily eschew.

To get a version of realism worth fighting for let us follow Goodman's hint and add kinds to our account of realism. This commits us to the view that there exists objective similarities between natural objects, i.e., there are natural kinds. Thus another version of realism stronger than OR2 but still weak in other respects is:

OR3: There are some kinds (such as electrons, 'flu viruses, kauri trees or galaxies) which exist in a suitably mind-independent manner (and which are open to scientific investigation).

The last two versions of ontological realism leave open, as they should, the possibility that there are mind-dependent objects or kinds which can be investigated by appropriate sciences e.g., after-images and natural languages are the province of psychology and linguistics. But on the definitions of scientific realism canvassed so far such items cannot be taken realistically. This raises a serious problem for all three versions of ontological realism since they require some account of what it is to 'exist in a suitably mind-independent manner'. No attempt will be made to give such an account here.[15] It is a fairly intuitive requirement which, however it is spelled out, is meant to preclude the possibility of some non-realist reconstrual of the items that are alleged to exist, for example, a phenomenalist reconstrual of objects or kinds (or, if the path of semantic ascent is followed, phenomenalist language reconstructions of talk of objects, or kinds). Given the current low success rate for the phenomenalist programme, all three versions of ontological realism seem to be immune to at least that kind of anti-realist reconstrual. The denial of OR3 yields at least the view that there are no mind-independent kinds, that is, there are no naturally occurring kinds but only the kinds that we construct as a result of our epistemic activities of classifying what objects there are. While this could still remain realist with respect to objects it is non-realist with respect to kinds.

OR2 or OR3 are supported by many philosophers including Devitt, Hacking, Cartwright, Ellis and Harré.[16] Hacking's crisp slogan "if you can spray them then they are real" captures the experimental scientists' practical sense of the reality of items which, though they cannot be observed or perhaps can never be observed directly, can be manipulated to produce observable effects. Once electrons were discovered they entered into a vast number of experiments such as Millikan's oil drop experiment; merely by spraying electrons onto oil drops an experimenter can watch the changing motions of the droplets. Hence the experimentalists' belief that they are working with really existing entities. Though causal manipulability may be a good test for whether or not an object exists it cannot serve as a useful criterion for ontological realism. There is no reason to suppose that all mind-independent entities can be manipulated to produce observable effects; neutrinos and tectonic plates exist but are hardly manipulable by us in experiments.

2.2. *Semantic Realisms*

Semantic formulations of realism involve reference to scientific theories; they concern either the truth-values of sentences of a theory or the referents of terms of a theory. Such formulations usually stand in contrast to instrumentalist interpretations of theories. In order to get instrumentalism off the ground a distinction needs to be drawn between observational ('O' for short) sentences and non-observational ('non-O' for short) sentences of a theory and between O-terms and non-O-terms of a theory. How these two distinctions are drawn is too large a topic for us to enter into here, but drawn they must be if there is to be instrumentalism at all. The distinctions may be grounded in epistemological considerations involving perception or they may be grounded in the theory of meaning. In the latter case we may understand O-terms and O-sentences as short for *old* terms and sentences in that their meaning has been fixed prior to their use in theory, while we may understand non-O terms and sentences as short for *new* terms and sentences in that their meaning has yet to be fixed in theory. In a nutshell, there are two varieties of instrumentalism. In one variety, the non-O sentences are regarded as non-propositional, i.e., they lack a truth-value. They are used as instruments or tools for constructing links between bunches of O-sentences (in some cases as inference rules linking observational premises with observational conclusions, i.e., predictions). A second variety of instrumentalism is based on some method of eliminating non-O-terms and sentences from a theory. One such "eliminativist instrumentalism" is due to Ramsey; all non-O terms are eliminated from the sentences in which they occur in favour of sentences with just quantifiers and O-terms.

In contrast to the truth-valueless version of instrumentalism Hellman[17] suggests a weak version of semantic realism:

SR1: Some non-O sentences have a truth value.

Thus some theoretical states of affairs make some theory true or false. However this is a quite weak version of semantic realism in which other theories can be given an instrumentalist interpretation; also it is compatible with the entire history of science yielding nothing but false theory. Stronger versions can be obtained by replacing 'some' by 'all' and by requiring that the non-O sentences have some degree of verisimilitude (assuming some appropriate account of verisimilitude). For those with qualms about how the O/non-O distinction can be drawn, Hellman suggests that realists should take their cue from their opponents and say that, given the way *they* draw the distinction, semantic realism can still be given some formulation:

SR2: There is no defensible demarcation between O and non-O sentences in science generally such that only the former have truth-value.

These, of course, are not the only ways in which realist and anti-realist theses may be defined in terms of the truth-values of sentences. One influential line is that of Dummett;[18] however even a cursory treatment of his views cannot be undertaken here.

Against those who would eliminate non-O terms are those who would keep them and take their designatory role quite seriously. Strong versions of semantic realism can be obtained by requiring not only that the non-O sentences have a truth value but also that most or all of the non-O terms have a designation. Such a view, and more besides, is held by Boyd, as reported by Putnam. Realism is regarded as an 'overarching empirical hypothesis' which claims:

SR3: (1) Terms in a mature science typically refer. (2) The laws of a theory belonging to a mature science are typically approximately true.[19]

Such a version of realism requires that mature scientific theories have got it nearly right in that the non-O sentences not only have a truth value but also have high verisimilitude. However it leaves open the possibility that our mature theories may, but only in part, be about things that do not exist, e.g., the term 'quark' may be found in the long run to have really lacked a designatum all along but this will not discredit the rest of particle physics.

All would be fine with the above if the notions of reference and truth upon which they depend were unproblematic. But notoriously they are not! Consider the relation of reference. The views of the early Putnam and Kripke were that there is a straightforward determinate relation of reference between names or kind terms and bits of the world. The later Putnam has come to reject this view which he regards as part of what he calls 'metaphysical realism', i.e., the view that there is a notion of THE WORLD which is independent of any representation of it.[20] If the views of the early Putnam on reference are intended in clause (1) of SR3, then SR3 is a thesis of metaphysical realism. It is part of Putnam's doctrine of internal realism that such determinate relations of reference are not available at all; thus SR3 could also be given an alternative 'internalist' reading. Thus the verdict on the viability of SR3 as an account of a realism hinges on issues elsewhere in the theory of reference and truth. Here is one instance where realist and non-realist stances, as Blackburn puts it, 'form obscure alliances with, and hostilities towards, various views of truth: correspondence, coherence, pragmatic, redundancy, semantic, theories'.[21] The same can be said about rival theories of reference as well.

2.3. *Epistemological Realisms*

For hard line realists epistemological varieties of realism are hardly worth the name. They claim that there is no privileged status to be accorded to the perceptual means whereby we acquire our theories of the world. Perception is simply part of the causal nexus of the realist's world which encompasses both perceiver and perceived; there are no lingering perceptual items (e.g., Lockean sensations, Humean impressions or sense-data) out of which to construct a 'veil of perception'. However Hellman suggests ways in which epistemic operators such as 'it is reasonable to believe that', 'there exists (good) evidence that', 'it is known that', 'it is certain that' or 'it is highly probably that' enter into the above formulations of both ontological realism and semantic realism to yield a range of epistemic variants on each.[22] Thus consider SR1. An epistemic version of this can be constructed thus:

ER1: It is reasonable to believe that (or there exists good evidence that, or it is known that, etc.) some non-O sentences have a truth value.

Given the variety of realisms already canvassed and the varying degrees of strength with which they can be formulated, adding any of the epistemic operators mentioned above proliferates the variety of realisms.

Realists tend to be guided by what our best current and most comprehensive theory tells us about the world even though, of course, it is neither a full nor infallible guide to what there is. Further, the entities postulated in our theories are often non-observable in some sense of that term (though what the realist may admit as an observable item may go well beyond what, say, an empiricist may admit).[23] Granted this we may formulate a strong version of epistemic realism along the following lines:

ER2: There is good evidence to the effect that theory T is our best tested most comprehensive theory; T tells us that entities E_1, E_2, ... , E_n exist and that there are laws L_1, L_2, ... , L_m; therefore it is reasonable to believe (a) that these entities exist and (b) that those laws are approximately true.

Entity realists of the kind described by OR2 or OR3 (such as Hacking or Cartwright) might accept some epistemic reformulation of their realism of the sort indicated in clause (a), but they would definitely reject clause (b). Those who hold not only some version of semantic realism but also some account of verisimilitude would be willing to countenance clause (b) as well, providing they also accept epistemic formulations of their realism. A quite strong form of realism is the claim that even our best scientific theory may fail to capture how the world is:

ER3: It is possible that in the ideal limit of scientific investigation our final scientific theory is false.

This is a mixed modal-epistemic claim about our application of the best canons of scientific method to ultimately yield a theory of the world.

The above is only a sketch of some of the varieties of realism that can be distinguished and expressed within the compass of single sentence formulations. Clearly more might be said to spell out these varieties in greater detail. Further, no arguments have been offered for or against any of these formulations; at best only a few hints have been given that some of them might be open to anti-realist construal and that more careful formulations might be needed. Moreover nothing has been said about the threat of underdetermination of theory for realism, the special problems of realism for theory in quantum mechanics and a host of other issues that concern realism. However the above will suffice for the purposes of this introductory survey. Consider, now, relativism and a similar three-fold way of formulating some of its varieties.

3. VARIETIES OF RELATIVISM

Protagoras' Measure Hypothesis was taken by Plato to be a relativization of either perceptions or truths to individual persons. Other relativizers could be considered, e.g., social classes, cultures, historical epochs, species, languages, categorical schemes, forms of life and frameworks or general theories or paradigms. The varieties of relativism could then be distinguished by means of distinctive relativizers. But they could also be distinguished by means of what is being relativized. Thus truth is often alleged by relativists to be relative to persons, classes, frameworks or whatever. This would yield a number of cases of what will be called 'semantic relativism'. In one of Plato's interpretations of Protagoras it is alleged that perceptual experiences are relative to persons; and sometimes the objects postulated in science are alleged to be relative to some general framework or paradigm. These would yield instances of what will be called 'ontological relativism'. Finally methods of investigation (as in science) and even rules of reasoning are alleged to be relative to cultures or to general frameworks or paradigms. This, as will be seen, turns out to be one variety of what will be called 'epistemological relativism'. Not all of these varieties of relativism are independent of one another but they turn out to be sufficiently distinctive to warrant separate classification. (There are non-cognitive versions of relativism to be found in ethics and aesthetics; these will not be considered here.)

3.1. *Ontological Relativisms*

Ontological relativism is the view that what exists, whether it be ordinary objects, facts, the entities postulated in science, etc., exists only relative to some relativizer, whether that be a person, a theory or whatever. One version of ontological relativism is given in the *Theaetetus* (160 B-C) where Socrates says:[24]

> Then my perception is true for me; for its object at any moment is my reality, and I am, as Protagoras says, a judge of what is for me, that it is and of what is not, that it is not.

Here the relativizer is an individual person. What is relativized may be either (1) a perceptual judgment which is claimed to be true for me — this yields a version of semantic relativism, or (2) a perception which is claimed to be real for me — this yields a version of ontological relativism. Plato does not always distinguish, as we would, between perceptual judgments and perceptions; thus the two versions of relativism are run together. Full semantic relativism arises when each of the judgments I make (these need not be restricted to only perceptual judgements) is alleged to be true for me. (Plato, in the first clause of the quotation makes the 'for me' relativizer quite explicit; however in his report of what Protagoras says he often omits it.) According to Protagoras not only are truths relative to me but also what is is relative to me, i.e., talk of 'my reality' becomes appropriate in this context. Since truth-makers are what make judgments true, then truth-makers are denizens of the world for me — or, more correctly, truth-makers for me. Thus semantic relativism of the true-for-me variety entails, in virtue of there being truth-makers for me, a version of ontological relativism.

A different kind of ontological relativism arises when perceptual experiences are the objects which, as Socrates puts it, are 'at any moment my reality'. Each person has his or her own perceptual experiences and these are not had by any other person. Moreover, for each person the perceptual experiences can change quickly over a short time. Phenomenalists would perhaps, agree with this much but this does not make them relativists. This raises a serious issue about what relativists intend by their doctrines. That perceptions are dependent on perceivers and that such dependence is always two-place or relational does not entail any substantive doctrine of relativism. We may say, as a *façon de parler*, that perceptions are relative to perceivers but this in itself yields no relativism. It is not obvious what the relativist ought to add to the view that perceptual experiences are perceiver-dependent; sometimes what they add is not even coherent. Consider adding the claim that each of us lives in a distinct but private world, a world relative to each person, which comprises only the flux, from moment to moment, of perceptual experiences. Expressing the matter this way gives the lie to the relativist position; there is at least a God-like viewpoint from which it can be said that *each* person is in what is for them a world of their own perceptual experiences. This raises questions concerning the overall coherence of the relativists' position; however the several cogent arguments against such relativism will not be pursued here.

Ontological relativism can come in other quite distinct varieties. The entities postulated in science are sometimes alleged to exist only relative to a sufficiently broad and deep scientific theory, or a paradigm or scientific framework; they do not exist in any independent manner. The following remarks of Kuhn seem to have a strong flavour about them of the relativity of ontology to a scientific theory:

> . . . I do not myself feel that I am a relativist. Nevertheless, there is another step, or kind of step, which many philosophers of science wish to take and which I refuse. They wish, that is, to compare theories as representations of nature, as statements about 'what is really out there.' Granting that neither theory of a historical pair is true, they nonetheless seek a sense in which the latter is a better approximation to the truth. I believe that nothing of that sort can be found.[25]

Clearly Kuhn is not a semantic realist of the strong SR3 variety. However he leaves it open as to whether, for theories taken separately, he is a weak semantic realist, say, of the SR1 variety, or whether he is an ontological realist of any sort. Some of these options are foreclosed when Kuhn rejects a correspondence theory of the sort realists require as part of their semantic realism (either strong or weak). Such a rejection does not entail relativism, of course, though the following remarks seem to point in that direction, despite the disclaimer at the end:

> There is, I think, no theory-independent way to reconstruct phrases like 'really there'; the notion of a match between the ontology of a theory and its "real" counterpart in nature now seems to me illusive in principle. . . . Though the temptation to describe that position as relativist is understandable, the description seems to me to be wrong. Conversely, if the position be relativism, I cannot see that the relativist loses anything needed to account for the nature and development of the sciences.[26]

Kuhn does not describe the kind of relativism to which he thinks he is not committed. Much uncertainty lies in the way in which the phrase 'really there' is said not to be reconstructable in a theory-independent way. In one sense Kuhn's remark is correct. If electrons are 'really there' then that we come to know this is wholly due to the strength of the evidence for electron theory and the experimental technologies based on that theory. But this is a point about our *coming to know* that electrons are really there. It is not a point about electrons being really there in the first place. It is this latter point that Kuhn seems to deny; electrons are not really there in any manner independent of electron theory. For an ontological realist (and other realists as well) electrons are really there independently of any theory. For an ontological relativist electrons are 'really there' only in virtue of electron theory (or some other theory which includes electron theory as part).

Ontological relativism sometimes goes hand in hand with an explicitly activist or constructivist account of the objects or kinds of object that are alleged to exist (albeit relatively). If we are active makers of our scientific theories then a relativist might want to extend our activities of construction if not to the objects then to the kinds of object theories are about. Of course a realist can well agree with the claim that we are active makers of our theories but reject, correctly, any implication that we are active makers of what our theories are about. Despite this, ontological relativity of kinds is a common claim made by sociologists of science. Some reject the view that there are any kinds, i.e., they reject OR3, or they reject an epistemic version of OR3. Thus Barnes and Edge tell us: "No body of knowledge, nor any part of one, can capture, or at least can be known to capture, *the* basic pattern or structure inherent in some aspect of the natural world". So where do the kinds come from that our theories are allegedly about? They go on to say:

> Nature can be patterned in different ways: it will tolerate many different orderings without protest. . . . No particular ordering is intrinsically preferable to all others, and accordingly none is self-sustaining. Specific orderings are constructed not revealed, invented not discovered, in sequences of activity which however attentive to experience and to formal consistency could have been otherwise . . . .[27]

Barnes and Edge call this view conventionalism. That label aside, the patternings discerned in nature are clearly an invention of ourselves in so far as we are active theorizers about the natural world, i.e., the authors support OR1 and OR2 but in rejecting OR3 support constructive ontological relativism for kinds.

Ontological relativism finds a natural home in the doctrines of many theorists of society, e.g., those who speak of the 'social construction of reality' where that reality concerns social facts, objects or kinds. What is meant by that phrase needs careful investigation from author to author. Suffice to mention here that the entities allegedly constructed are not *sui generis* but are held to be dependent, in some sense, on other entities. Some might treat their dependence as simply one way of talking about their relative existence in which case the variety of relativism alleged is only as problematic as the notion of dependence advocated. Others might wish to hold a more full-blooded variety of ontological relativism for social entities in which case much work needs to be done to distinguish this kind of relativism from the former weaker sort (and perhaps even more work to make it a coherent form of relativism). Such constructivism finds its way into the sociology of science yielding some of the more implausible theses proclaimed in that area. Thus Collins maintains that the school of sociology with which he is associated "... embraces an explicit relativism in which the natural world has a small or non-existent role in the construction of scientific knowledge".[28] This is relativism with a vengance. There is talk of a natural world but such realism of the OR1 kind plays no significant role; rather, what is constructed is of prime significance, but this has almost nothing to do with the natural world!

3.2. *Semantic Relativisms*

That truth, and falsity, are relative is perhaps the commonest form of relativism. Truth and falsity have been claimed to be relativizable to a host of items from individuals to cultures and frameworks. What is relativized is variously sentences, statements, judgements or beliefs. On some but not all versions of relative truth it is claimed that if p is true for x then there is something that makes p true for x, i.e., there is a truth-maker for p which is part of the relativized world for x. Less full-blooded varieties of truth relativism can be obtained by uniting relativism with some non-correspondence theories of truth, e.g., coherence or pragmatic theories. However one must then be careful to distinguish such a relativism from other philosophical positions, for example, pragmatism, which are on the face of it non-relativist.

Semanticists often speak of 'truth in a model', 'truth under some interpretation' or 'truth in a possible world'. Tarski's Convention T also refers to truth in a language *L* thus: '*s*' is true in *L* if and only if *p* (where '*s*' is a structural-descriptive name of a sentence and '*p*' is the translation into the metalanguage of that sentence). Are any of these relativistic notions of truth? Clearly not all relational notions of truth need commit one to any substantive doctrine of relativism. Thus consider Tarski's project. This was to formulate definitions of truth for various formal languages in higher-order metalanguages. This project has been extended by Davidson and others to include natural languages. Both involve an attempt to provide a definition of 'true in *L*' where for '*L*' can be substituted some formal or natural language. Such a project does not involve one in any form of philosophical relativism despite the relational character of what is being defined. And the same can be said for talk of 'truth in some model' or 'truth under an interpretation' or 'truth in a possible world'. For example, truth in an interpretation merely concerns the way in which non-logical terms are assigned their extensions and how these extensions are related to one another. Relational these semantic notions of truth may be, but no doctrine of relativism flows from this.[29]

The Tarski view of truth can cast light on some of the darker aspects of relativized truth. As Tarski envisages,[30] the very same sentence '*s*' can, in one language, be true while in another be false, and in yet another be a meaningless expression (this being the more likely possibility). None of this threatens contradiction or pushes us in the direction of relativism. Again a case can be made for the same sentence '*s*' in the same language *L* being ascribed different interpretations and thus having different truth-values. To use Newton-Smith's example, the sentence 'grass is good to smoke' is false in the idolect of farmers but not of some members of communes.[31] For a substantive version of relativism we would require either that the very same sentence *s* under the same interpretation (i.e., it would have the same 'meaning') be true in *L* for some speakers and false for others, or if two languages *L* and *L′* are considered and if a sentence *s* in *L* translates into sentence *s′* in *L′* according to some translation manual, then *s* is true in *L* while *s′* is false in *L′*. Such a conception of relativism enables us to say that *s* is true relative to one group of speakers, *G*, but false for another group of speakers, *G′*. (The above can easily be adapted to other relativizers such as cultures or frameworks or whatever.) Newton-Smith argues that such a conception of relativism is incoherent.[32] What is it that is true for *G* but false for *G′*? To avoid trivial semantic relativism of the 'grass is good to smoke' variety we have to focus on sentences under a given interpretation, i.e., propositions. But since propositions are individuated by their truth-conditions then in the substantive version of relativism specified above the very same truth- conditions make a proposition true and make it false. The prospects for a viable truth relativism are grim indeed!

One other variety of semantic relativism concerns the designations of terms. Quine argues in 'Ontological Relativity' that the inscrutability of reference that infects the general terms in a language can only be overcome, relatively, by recourse to a background language:

> When we ask, "Does 'rabbit' really refer to rabbits?" someone can counter with the question "Refer to rabbits in what sense of 'rabbits'?" thus launching a regress; and we need the background language to regress into. The background language gives the query sense if only relative sense; sense relative in turn to it, this background language.[33]

The regress of background languages only stops, as Quine claims, 'by acquiescing in our mother tongue and taking its words at face value'. Thus the reference of terms like 'rabbit' is fixed not absolutely but relatively.

Incommensurability can, on many occasions, be understood as a version of semantic relativism. Because truth is allegedly relative to a theory the truth and falsity contents of many pairs of theories are incomparable. Again, because the terms of a theory have their referents fixed relative to that theory then in sufficiently different theoretical contexts the same term will have different referents. Thus Kuhn declares:

> In the transition from one theory to the next words change their meanings or conditions of applicability in subtle ways. Though most of the same signs are used before and after a revolution — e.g., force, mass, elements, compound, cell — the ways in which some of them attach to nature has somehow changed. Successive theories are thus, we say, incommensurable.[34]

Such incommensurability is grist to the relativist's mill and has excited much critical comment more of which appears in this volume in the papers by Kroon and Oddie.

3.3. *Epistemological Relativisms*

Epistemological relativism can arise in a number of ways. First, epistemic notions may be explicitly included in statements of relativism;

thus it may be alleged that what is known or believed is relative to a person, culture, framework or whatever. Here relativists adapt terms like 'know' and 'believe' to their own purposes using them in ways unfamiliar to traditional epistemologists. For traditionalists, '*x* believes that *p*' can be true while '*p*' is either true or false. For many relativists such a notion of belief is not available; they assimilate '*x* believes that *p*' to '*p* is true for *x*' thereby using 'belief', or 'true', in a quite non-standard sense. Similarly, if what is known is relative to some *x* then knowledge must be understood in the context of a relative and not an absolute sense of truth. Traditionalists would construe talk of knowledge in such contexts as a misleading way of talking of belief once more — but this would not be true to the relativists' intentions.

Second, epistemic relativism arises naturally in certain views about the nature of perceptual reports. Since the traditional foundationalist view of epistemology has come under severe attack some have drawn the conclusion that, in the absence of any firm foundation for our knowledge claims in perceptual reports, a relativist view of observation in science is the appropriate stance to adopt, i.e., there is incommensurability at the observational level. Such are the views of Kuhn, Feyerabend and many others who would not be relativists in other respects. Certainly Hanson's ruminations over the question 'Do Kepler and Tycho Brahe see the same thing in the east at dawn?'[35] and his and Kuhn's ruminations over the puzzles raised by changing aspects diagrams have led a great many to take a relativist stance with respect to observation in science. The catch-phrase is that "all observations are theory laden" from which it is commonly inferred that scientific observations are relative to some theory. The issues raised by this variety of epistemic relativism are too complex to be gone into here, especially some of the cogent objections that have been raised against it.[36]

A final kind of epistemic relativism arises with the claim that there is no theory of scientific method (or a methodology in any form of inquiry), or there are no rules of reasoning, especially evidential rules, which can be accorded a special or privileged status over and above any other method or set of rules of reasoning. The procedures whereby we establish any belief are not themselves absolute in any sense. It is not that there are several methodologies or sets of rules tied for first place as our privileged canons of reasoning which when applied to our theories or beliefs yield, on the whole, comparable rankings of these theories and beliefs as better or worse. Rather the different sets of methodologies and rules yield distinctly different rankings; as a consequence no ranking can be declared superior to another in any sense. Relativists commonly claim that methods and rules are as variable with respect to times, persons, cultures and frameworks as the very beliefs they establish or between which they are used to adjudicate. They admit that there can be local adjudication of beliefs in a given context, i.e., within a culture or framework beliefs can be ranked according to their acceptability or unacceptability. Rather, there are no global overarching canons of method or of reasoning applicable across all cultures, frameworks or whatever. Given the pervasiveness of this kind of epistemic relativism it perhaps deserves a name of its own – methodological relativism. The darling of methodological relativists must be Feyerabend who, in his ultra-anarchistic moments, has declared that in matters methodological "anything goes". Or if this is too strong a claim even for relativists, there are the more muted claims that many rules of scientific method are, in some sense, context-dependent, and that there is a range of traditions and practices none of which can be claimed superior to any others.[37] Even more muted are Hacking's considerations[38] on behalf of different 'styles of reasoning' for which he appropriates the term 'relativism'. However his reasons for using the term have more to do with the lack of a foundation for a style of reasoning than with the reasons commonly advanced by relativists for methodological relativism. We should not assume that any anti-foundational view of methodology entails the strong version of methodological relativism just cited. Even more mild is Doppelt's espousal of a modicum of relativity for theories of scientific rationality.[39]

One can be a methodological relativist without being an ontological relativist at the same time, e.g., Hacking who, as was noted, is an ontological realist. Some sociologists of knowledge also occupy a similar position. Thus Barnes and Bloor openly talk of an independent reality when they say: "The general conclusion is that reality is, after all, a common factor in all the vastly different cognitive responses that men produce to it".[40] Their brand of relativism is strictly methodological: "For the relativist there is no sense to be attached to the idea that some standards or beliefs are really rational as distinct from merely locally accepted as such".[41] As a consequence they maintain "'Evidencing reasons', then, are a prime target for sociological enquiry and explanation".[42] Their mentor has been Wittgenstein who proclaimed: "All testing, all confirmation and disconfirmation of a hypothesis takes place already within a system".[43] Talk of 'a system' is vague, but, as he suggests in a following paragraph,[44] there is propriety in talk of 'thinking in our system and of rivals to our system which we would find intellectually very distant'. Thus rival systems yield rival test results — a consequence a methodological relativist would welcome.

Given the variety of relativisms no one argument can show them all to be defective — if that is what they are. Apart from the few difficulties raised above no attempt will be made in this introduction to critically evaluate arguments for or against the various kinds of relativism. Truth relativism, since it plays a central role in relativism has been the most susceptible, and the most vulnerable, to attack. Plato advanced several considerations against it in the *Theaetetus*, the most well known being the self-refutation.[45] And, as was mentioned in section 1, a large number of contemporary philosophers have continued the debate.[46] In the next section the role of relativism in the sociology of science will be considered as it arises in the work of a few of its many defenders currently working in that field. A few problems will be raised concerning their formulations of relativism but, again, no extended critique will be offered of these formulations — that can be found in a number of the papers collected in this volume.

4. RELATIVISM AND THE SOCIOLOGY OF SCIENCE

Some variety of relativism usually accompanies doctrines in the sociology of science though there is no reason why this should be so. Sociologists of belief (or knowledge as they prefer to say, though this can be seriously misleading) claim that the wide variety of beliefs that people hold on any topic is one important reason for relativism. Thus Barnes and Bloor declare: "Our claim is that relativism is essential to all those disciplines such as anthropology, sociology, the history of institutions and ideas, and even cognitive psychology, which account for the diversity of systems of knowledge, their distribution and the manner of their change".[47] That there are varying beliefs held on any topic may be an impetus to theorizing in the sociology of belief but this is neither necessary nor sufficient for relativism in the sociology of belief. It is not sufficient for clearly non-relativists can, and do, try to account for the varieties of belief as well. Nor is it necessary. Suppose all of humanity were to hold the same beliefs on any given topic. Then the sociologist of belief would still have two tasks to perform, viz., to give an account of why that particular set of beliefs was held on a given topic and to give an account of why one and only one set of beliefs was held by all people. An all-encompassing totalitarian belief system would be as amenable to sociological investigation (whether relativist or not) as are the wide variety of beliefs actually encountered.

Continuing to take Barnes and Bloor as the spokesmen for one brand of relativism in the sociology of science we find them adding: "The simple starting point of relativist doctrines is (i) the observation that beliefs on a certain topic vary, and (ii) the conviction that which of these beliefs is found in a given context depends on, or is relative to, the circumstance of the users".[48] Since (i) is otiose what of (ii)? (ii) is a schematic meta-sociological claim that particular sociological investigations will, hopefully, exemplify. It says that for any individual x in some circumstance or context (call this 'C_x') and for any belief B that x holds then that x believes B depends on C_x (what counts as x's circumstance being left open). The relationship of dependence is not spelled out in this schema. It may be counterfactual dependence as in 'if x were in circumstance C then x would believe that B'; more often than not the authors speak of C_x causing x's believing that B. However the relation of dependence is interpreted, a point made in previous sections still holds. Relations of dependence do not entail any substantive doctrine of relativism even though it is a *façon de parler* to speak, as do Barnes and Bloor, of beliefs being relative to the believer's circumstance. Some other distinctive thesis needs to be added by relativists to claims about the dependence of belief on circumstance. What this might be remains obscure. Certainly Barnes and Bloor explicitly wish to avoid any kind of truth relativism as the differentia in this case.[49] So even a non-relativist sociologist of belief could embrace (ii) as a meta-sociological thesis about belief dependence (even though what counts as C_x has been left entirely open).

An equivalence postulate is the third important feature of relativism, say Barnes and Bloor, and they express it thus:

> Our equivalence postulate is that all beliefs are on a par with one another with respect to the causes of their credibility. . . . The position we shall defend is that the incidence of all beliefs without exception calls for empirical investigation and must be accounted for by finding the specific, local causes of this credibility. This means that regardless of whether the sociologist evaluates a belief as true or rational, or as false or irrational, he must search for the causes of its credibility.[50]

This is an expression of Bloor's earlier impartiality criterion for his strong programme in the sociology of knowledge. However if we construe the talk in (ii) of dependence as causal dependence then the equivalence postulate is merely a consequence of (ii) since (ii) is universally quantified with respect to *B* irrespective of whether these beliefs are true, false, rational or irrational. Thus the equivalence postulate adds nothing to (ii) in the way of a more substantive characterization of relativism; it merely emphasizes that (ii) concerns *all* beliefs.

One way in which (ii) might be thought to entail relativism is by placing scientific beliefs beyond the pale of rational assessment. But this is not so; (ii) can be construed in a way compatible with rational assessment of beliefs. So far no restriction has been placed on what C_x may be. Sociologists would want to include not only *x*'s political, social and cultural circumstances but also *x*'s biological and intellectual circumstances and *x*'s cognitive capacities. Thus as Barnes and Bloor indicate, in hunting for the particular circumstances upon which beliefs depend a sociologist may sensibly ask "... if a belief is part of the routine cognitive and technical competences handed down from generation to generation". Further they add: "[Sociologists] simply investigate the contingent determinants of belief and reasoning without regard to whether the beliefs are true or the inferences are rational".[51] Thus a host of rules of reasoning no matter whether they are valid or invalid, from *Modus Ponens* to the fallacy of affirming the consequent or the gamblers' fallacy, may be part of a person's routine cognitive competences which are socially transmitted from generation to generation.[52] Since these, too, are part of C_x we have as an instance of (ii): *x*'s believing *B* is dependent (causally or not) on a particular circumstance of *x*, *viz.*, *x*'s having rule *R* (valid or not) as part of *x*'s cognitive competences. Most philosophers do subscribe to the view that *acts of believing*, as distinct from *belief contents* or *propositions*, have causes, i.e., *x*'s believing *B* is not something which is outside the mind-body causal nexus or which is without any causal antecedents at all.[53] Amongst these causes can be include such things as *x*'s having a rule *R* as part of *x*'s cognitive competences. So, given a wide enough construal of what may fall under the extension of C_x, there is nothing about (ii) which need perturb the philosopher who holds the quite weak claim that acts of *believing* (as distinct from *belief contents*) have causes. In fact (ii) says little more than this unless C_x is restricted in some way to a particular kind of cause. If C_x were restricted to only social contexts then (ii) would be a clearly false empirical claim. Thus (ii) broadly construed entails no relativism — at worst it is relativism in name only. Since there are no grounds for conflict between sociologists and philosophers concerning (ii) then the several excellent sociological studies carried out in the light of (ii) ought to be welcomed as a significant contribution to our knowledge concerning the determinants of belief, including scientific beliefs.

One important issue, passed over so far, is whether rule *R* is valid or not. So far only *x*'s *holding* rule *R* has been alleged to be a cause of *x*'s believing *B*. This factual matter is quite independent of the evaluative question of whether *R* is a good rule or not; *R* could be the acceptable rule of *Modus Ponens* or it could be the unacceptable gambler's fallacy. The relativist ingredient in the Barnes and Bloor position turns out to have nothing to do with (ii). Rather their methodological relativism emerges as an independent doctrine when they declare: "For the relativist there is no sense attached to the idea that some standards or beliefs are really rational as distinct from merely locally accepted as such".[54] Now it has been admitted that as far as local acceptance is concerned *x*'s accepting rule *R* (valid or not) can be a cause of *x*'s believing that *B*. This bears not at all on the truth or falsity of *B* or the rationality or irrationality of believing *B* or the validity or invalidity of rule *R*. Thus in one sense philosophers could agree with Barnes' and Bloor's claim that "'evidencing reasons' are a prime target for sociological enquiry and explanation"[55] even though they might demur from part of this claim, *viz*., that the enquiry would be sociological only and not have elements from other areas of inquiry such as psychology. However, philosophers would disagree with the claim understood in the sense that sociology (or any other empirical science) could ever answer the evaluative question of whether some alleged 'evidencing reasons' were good or bad reasons.

Barnes' and Bloor's relativism emerges as a scepticism about whether or not we can ever establish that any rule of reasoning has any rational backing:

> [Justifications] are circular because they appeal to the very principles of inference that are in question. In this respect the justification of deduction is in the same predicament as the justification of induction which tacitly makes inductive moves by appealing to the fact that induction 'works'. Our two basic modes of reasoning are in an equally hopeless state with regard to their rational justification.[56]

Given such scepticism it is clear that not even sociology should attempt to embark on the otherwise allegedly futile project of discovering what are the good and the bad rules of reasoning. The sociologist merely tells us which rules are accepted or rejected by a culture or by a believer in the culture. (The same must hold for the sociologists' enterprise and the grounds for holding (ii); this Barnes and Bloor would accept since they do espouse a reflexivity principle to the effect that principles of sociology, including (ii) itself, are not exempt from the scope of (ii).) Their scepticism about the rational justification of rules of reasoning backs their methodological relativism. Countering their scepticism is a vast project not to be undertaken here. Suffice to mention that the reason the authors offer for their scepticism about justifications in reasoning is that they are circular. However not all attempts at such justifications are circular. Since Hume first pointed out the circularity in one attempt to justify induction other attempts have studiously avoided the circularity change. At the same time they have considerably refined our intuitions about what is meant by 'rational justification' in such contexts.

So far relativism in the sociology of science has been discussed only with respect to Barnes' and Bloor's succinct statement of their views. However their position is representative of many working in the field. Relativism of other forms has other advocates as well, it being evident by now that relativism is as protean a set of doctrines as realism — even within the sociology of science. In picking one other view we find, for example, that Collins advocates what he calls the 'empirical programme of relativism'; this differs in some respects from Bloor's strong programme though (ii) is an important ingredient in both.[57] Relativism emerges in three ways in Collins' programme. First, there is an interpretive flexibility about experiments the final interpretation of which is a matter of social negotiation between scientists. Second, there are limitations to this flexibility which allow controversies in this area to come to an end. Finally, the first two features are related to the wider political and social structure.[58] Collins lists many studies which exemplify his programme. But the ways in which his programme supports a substantive variety of relativism are unclear. Scepticism seems once more to fuel relativism:

> We need to suspend that commonsense and philosophical view of science that gives us certainly about scientists' dealings with nature. What we need is radical uncertainty about how things about nature are known. This radical uncertainty is relativism.[59]

That science leads to certainty has hardly been a widely held view this century. In addition not all responses to the absence of certainty have lead to a radical scepticism 'about how things about nature are known'. But in the case of many sociologists of science such scepticism has been taken to be a green light for relativism about methods, reasoning, observation and ontology in science. However scepticism is neither the same as relativism nor does it entail it. Nor is it obvious that we need a heavy dose of either scepticism or relativism to do science, and, in particular, to do that science which takes other sciences (and itself) as its object of study. Given these disclaimers note that it does not follow that there is no viable study of the dependence of scientists' beliefs on particular circumstances (whatever they be) of the scientists.[60] What remains obscure is what sceptical relativism[61] adds over and above these dependence claims that can be established by empirical research in the sociology of science.

5. SUMMARY OF THE CONTRIBUTIONS

In the first paper in this collection David Papineau asks whether the new sociology of science, which investigates the influences which shape the very content of scientific theories, actually discredits science. In arguing that it need not, Papineau invites us to give up our traditional Cartesian epistemology which would discredit the new sociology and its picture of science and adopt a naturalized theory of epistemology which would not. Cartesians require us to consciously base our knowledge claims on logically secure arguments which stem ultimately from premises concerning our privileged awareness of our own mental states. In marked contrast the naturalized view requires us to ensure that all our beliefs come from belief-forming processes that are reliable for their truth. Justification for a belief is, then, not of the Cartesian variety; rather the justification is to be found in the reliability of the belief-forming process no matter whether the processes impinge on people individually or in groups (in which case the beliefs arise, in part, from reliable social processes). The new sociology of science highlights the social context in which scientists acquire their beliefs thereby playing down the conscious rational factors in belief formation to the dismay of the Cartesian epistemologist. However Papineau argues that even if we grant the correctness of the case studies of the new sociology this would not show that scientific practice is not generally reliable for producing true theories. The reliabilist epistemological setting for the new sociology of science restores the credit we have traditionally accorded science while at the same time rescuing the new sociology from some of its more extraordinary relativist claims about epistemology.

John Fox's paper is a critique of ethnomethodological studies of science. In particular he discusses the work of Garfinkel, Lynch and Livingston on the optical discovery of pulsars, and Latour and Woolgar's study of the Salk laboratories. The tables are turned on the ethnomethodologists. They allege that natural scientists and philosophers misunderstand the nature of their own enterprise when they claim to make discoveries about independent objective matters of fact. In response Fox argues that ethnomethodologists, given their own stance, misunderstand their own enterprise in that they have not discovered or revealed any independent or objective matters of fact about scientists and their activities. Several notions central to the ethnomethodologists' account of science are not particularly transparent. They construe terms such as 'fact' and 'realism' in their own way; they speak, for example, of the optically discovered pulsar as a "galilean" object which is a cultural object and not a physical or natural object thus rejecting a plausible realism with respect to pulsars; they talk of the 'social construction' of facts and objects. Fox provides a detailed critique of what goes wrong in the ethnomethodologists' use of such notions and shows why discourse analysis, currently employed in sociological accounts of science, is falsely based in theses derived from the more extreme wing of the ethnomethodologists' programme.

Philip Pettit argues that the four central tenets of Bloor's strong programme in the sociology of science can be understood in a way that does not entail relativism, particularly methodological relativism. Rather, the relativism associated with the programme is best explained by an independent commitment to a kind of conservatism. Conservatism arises because upholders of the strong programme maintain that no one ought to be disturbed by the revelation that their beliefs have a particular cause even if that cause should be such as to discredit those beliefs. The sociologists' enterprise is entirely free of any evaluation of beliefs; in particular, the enterprise admits no grounds on which one may challenge or change the beliefs of those under sociological investigation. In such a value-free atmosphere the strong programme can have no critical function. Pettit argues that if such conservatism is abandoned then an adherent of the strong programme could permit it to have a critical role. The strong programme does not permit the rational assessment of a belief to rule out a causal explanation of that belief; however the causal explanation of why a belief is held does not rule out, and could lead to, a critical reconsideration of the adherence to that belief. In the light of Pettit's critical reading of the strong programme relativism can be seen to be otiose.

Cliff Hooker and Kai Hahlweg argue that evolutionary epistemology does not lead to relativism. They are aware of the varieties of relativism that can be found and sketch nine considerations on the basis of which arguments for relativism might be advanced, the last two being arguments from biology. Evolutionary epistemology understood as a form of relativism is alleged to commit us to the following: (a) since theories are viewed as well-adapted when they are empirically adequate (i.e., they are well-adapted to their environment of application) there is no notion of theoretical progress available within evolutionary epistemology; (b) there is a deep-seated species relativism concerning knowledge since cognitive capacities develop as a series of adaptations based on an initial genetic inheritance and a contingent series of environments through which the species has evolved. The authors outline ways in which the arguments for relativism can be disarmed, especially the two from biology. They place the theory of evolution and evolutionary epistemology within a strongly realist framework for science. Species adapt physically to their environments, and some are able to adapt cognitively. The authors sketch a theory of changing cognitive adaptation based on the dynamic relationship between the creation and solution of problems in both theoretical and practical contents. It is only the realist and not the relativist framework that enables the theory of cognitive dynamics to do justice to the full range of theoretical, practical and technological developments to be found in our evolving epistemological framework.

Larry Laudan's contribution is cast in the form of a dialogue. A relativist and a non-relativist argue the virtues of a version of methodological relativism which alleges that we can never be in a position to claim that one theory is superior in any way to another. A number of arguments for and against such a relativism are aired by the two disputants. The first is an argument for relativism from the underdetermination of theory by evidence. The next is an argument based on the ambiguity or amorphousness of the rules that have been proposed for theory choice. The claim by Popper and Lakatos that the rules of theory choice are simply conventions is countered by the non-relativist who says that rules can be judged by the way in which they promote our cognitive aims. The relativist responds with an argument based on axiological relativity, viz., there is no non-arbitrary ground for claiming that any aim for inquiry is better than any other. By a series of adroit manoeuvres the non-relativist gets the relativist to admit that there are in fact good grounds upon which we can critically evaluate the aims of any inquiry. The upshot is that there are a number of grounds upon which consensus about theory-choice can be achieved that undercut considerations advanced in favour of methodological relativism.

Incommensurability is commonly cited as one ground for either ontological or semantic relativism in science. Fred Kroon investigates the credentials of the claim that terms which can occur in two or more theoretical contexts are referentially incommensurable. In particular, he examines the view that a descriptivist account of the meaning and reference of terms leads to an anti-realist view of science while a theory of direct reference such as the causal theory (advocated by the early Putnam and by Kripke) leads to a realist view of theories. Kroon points out the extent to which several philosophers, for example, Kuhn, Feyerabend and Rorty, have for various purposes deployed "descriptivism-to-anti-realism" arguments. Such arguments are shown to be defective, but it does not follow that a causal theory of reference is the appropriate view for a realist to adopt. Kroon in fact rejects both descriptivism in its classical form and any direct reference theory for theoretical terms in science, and advocates as the appropriate theory of reference-fixing a modified causal descriptivist account which is not inherently anti-realist. In order to conveniently marry a theory of reference with realism Kroon considers a version of epistemic realism which requires that there be informational states available to humans which certify that particular theoretical facts obtain. For a person to use a term to refer to some item more is needed than the bare causal connection between term and item required by direct reference theories; in contrast, less is needed than the full battery of descriptions required by classical descriptivism to be gathered from the term's theoretical context. In Kroon's theory of causal descriptivism what is important is the epistemic warrant that referrers possess if a term they use is to refer successfully to some item. Thus the epistemic conditions for the use of terms to refer, when united with an epistemic version of realism, enable us to avoid both massive term incommensurability of the sort that follows from classical descriptivism and complete referential success for our terms of the sort that follows from direct reference theories.

Incommensurability is tackled from another direction in the paper by Graham Oddie. He argues that anti-realists need not eschew a large measure of commensurability for theories and that arguments purporting to show that they should are mistaken. By the same token realists are committed to more incommensurability than they might have thought initially. Oddie embraces a strong version of semantic realism involving a notion of verisimilitude which is verification transcendent; his realism also has epistemic overtones to the effect that one of our aims in inquiry is to make progress in the sense of achieving greater verisimilitude. Incommensurability comes in two varieties, logical, in which two theories cannot be compared for their relative verisimilitude, and epistemic, in which there are no evidential criteria for judging relative verisimilitude. It gets its purchase from doctrines about the meanings of terms in theories, especially the view that a theory partially defines its own terms. Unlike Kroon, Oddie does not discuss theories of reference. Rather he considers a Ramseyfied method of picking out the analytic components of theories advocated by the later Carnap as the way of partially defining theoretical terms. What Oddie points out is that this proposal of Carnap's, unlike some earlier ones, does not entail meaning-variance for rival theories. In fact this proposal is compatible with a large degree of commensurability between theories — even in cases where meaning variance arises. Moreover, since Carnap's proposal is in a certain sense anti-realist, it follows that one brand of anti-realism is compatible with commensurability; thus anti-realism with respect to theories does not lead automatically to incommensurability as many have thought. Realism of a robust sort requires a considerable degree of commensurability. However Oddie argues that even robust realism must countenance some degree of incommensurability; in particular, since realists maintain that the factual content of a theory goes beyond its observational content then it will commonly be the case that different theories in the same domain will yield considerable incommensurability of a sort that an anti-realist manages to avoid.

Greg Currie raises a question concerning the status of social entities such as classes, banks, laboratories, etc. He argues that the variety of ontological realism we commonly adopt with respect to non-social objects, kinds or facts is not extendable to social objects, kinds or facts.

The reasons for this do not stem from a general anti-realist view of our knowledge of the world, but have to do with reasons peculiar to the social sciences themselves. What natural kinds there are and what laws of nature hold are independent of human thought and action; we can be ontological realists with respect to these kinds and laws. In contrast, social entities depend in some sense for their nature and their existence on human thought and action; thus we cannot be full-blooded ontological realists with respect to these entities but must be anti-realists instead. Currie does not espouse any form of ontological relativism with respect to social entities. Rather he spells out the relation of dependence that holds between social entities and human thought and action as one of supervenience. However social entities, even though they are supervenient on our thought and action, still possess sufficient independence and resilience to be things we find hard to change, if we can change them at all, or to be things which are not easily knowable by us. Currie also argues that there can be laws governing the behaviour of natural kinds but there can be no social laws (though there may be social regularities). This has an important consequence for methodology in the social sciences that distinguishes it from methodology in the natural sciences: if there are no social laws then it cannot be part of social science methodology to aim to discover them. Given Currie's analysis of social entities sociologists of science would have to reconsider the kinds of relations they alleged to hold between the beliefs of individual scientists and the social and other contexts in which those scientists exist.

Alan Musgrave sets himself the task of discovering the 'ultimate argument' for scientific realism. Roughly put, the argument is that realism unlike, say, phenomenalism, does not make the success of science a miracle, or that realism is the best explanation of the success of science. But neither of these claims look like arguments. The success of science, Musgrave contends, is best expressed by the following claim: a scientific theory T makes surprisingly novel predictions which turn out to be true. Let this be an explanandum of which realism and its rivals are required to provide an explanation. Musgrave suggests that realism's explanation comprises the two premises: (i) T is true and (ii) T yields surprisingly novel predictions. From these two premises the explanandum follows deductively. The realism that explains the novel predictive success of theories is a semantic variety of realism, *viz.*, the claim that T is true — or has high verisimilitude, a qualification that can be set aside for our purposes. The next question that can be asked is whether realism is the best explanation that can be offered for novel predictive success in science. Musgrave looks at various forms of inference to the best explanation and finds them all wanting. So he replaces them by a deductive argument which he thinks best captures the intentions of "explanationists". The task now is to evaluate rival explananda for the explanandum above. Rival candidates for (i) include: *T* is empirically adequate (Van Fraassen); *T* correctly solves all its empirical problems (Laudan); the world is *as if T* were true (surrealism). Musgrave concludes that realism provides a better explanation than any of its three rivals for their common explanandum. In what way is it better? The realist explanans is at least wider in scope than any of its rivals; moreover the rivals explain, in so far as they do, why *some* empirical consequences of a theory are true by claiming in effect that *all* are. Finally the realist can appeal to truth and reference as the grounds for a theory's success. If the anti-realist remains unmoved by this consideration then the realist can fall back on the previous two claims which turn on broadness of scope as an explanatory virtue. However for anti-realists who doubt that science itself is explanatory even this much will fail to convince them of the force of Musgrave's reconstrual of the ultimate argument for realism.

Richard Sylvan's philosophical stance is neither realist nor anti-realist (in the sense of, say verificationism). Nor is it relativist. It is pluralist. Versions of ontological realism require some notion of the mind-independence of what exists and thus stand opposed to verificationism. Realism also requires that there exists a unique world which fixes one and only one theory as that which corresponds to reality. Both pluralism and relativism reject the uniqueness requirement; characteristically pluralism and ontological relativism are many worlds doctrines. However pluralism and relativism part company on methodological grounds. Methodological relativism does not admit that there can be any overall ranking of theories or any criteria of choice while pluralism admits both ranking and choice. Perhaps many relativists are pluralists at heart but they state their philosophical views in the misleading and at times incoherent language of traditional relativism. Sylvan makes a clear distinction between the two and openly embraces a full and radical version of pluralism. Central to Sylvan's position is the view that there is more than one actual world. Many worlds doctrines can be found in Plato, Leibniz, Popper and in the modal realism of Lewis but none of these maintain the view that there are many actual worlds. Nor is the many actual worlds doctrine to be confused either with the claim that there can be many different theories of one actual world or with the claim that the many different worlds are merely human projections (either projected individually or socially as in doctrines of the 'social construction of reality'). In this respect Sylvan is thoroughly realist (in one sense of this term) with respect to the plurality of actual worlds. On the methodological front Sylvan also rejects the claim that there is one correct logic or one correct theory of method; there are in fact rival correct systems of logic and method. None of this, Sylvan argues, is incompatible with some kind of convergence of our knowledge as science advances; nor is it incompatible with the pruning of inadmissible theories or with the ranking of theories. What is rejected is the uniqueness claim concerning one correct logic, method, theory or world. There is much that relativists might find congenial in Sylvan's position; however they will not find any of the traditional forms of relativism or its defences in his paper.

University of Auckland

NOTES

[1] Lenin (1908), Chapter 2, Section 3 'Does Objective Truth Exist' and Section 4 'Absolute and Relative Truth, or the Eclecticism of Engels as discovered by A. Bogdanov'.

[2] Husserl (1970), Chapter 7 'Psychologism as a Sceptical Relativism', pp. 135—46.

[3] Passmore (1961), Chapter 4, especially pp. 64—9.

[4] Davidson (1984), Essay 13 'On the Very Idea of a Conceptual Scheme'.

[5] Putnam (1981); see the index under 'relativism'.

[6] Newton-Smith (1982).

[7] Burnyeat (1976).

[8] Rorty (1982), Chapter 9 'Pragmatism, Relativism and Irrationalism', even though some of his other pronouncements elsewhere have a relativist ring to them.

[9] Siegel (1986).

[10] On quite different grounds one could include here such people as Feyerabend (1978), pp. 9 and 28 which openly espouse Protagorean relativism, Meiland in a series of papers (e.g. Meiland (1977) and (1979) (which are criticized in Siegel (1986)) and Margolis (1986). See also the issue of *The Monist*, Vol. 67, No. 3, 1984, entitled 'Is Relativism Defensible?' in which various defences and objections are raised, and the collection Meiland and Krausz (1982).

[11] See Putnam's comments on p. 288 of Putnam (1983); see also pp. 234—8 for further criticisms of relativism.

[12] See Hellman (1983); for another three-fold distinction, see also Merrill (1980) who, however, presents his distinctions somewhat differently.
[13] Devitt (1984), p. 15.
[14] Goodman (1978), p. 20.
[15] For a fuller discussion of the issues involved here see Sober (1982), Section 2 of Hellman (1983) and Devitt (1984), Sections 2.2 to 2.5.
[16] See Hacking (1983), especially Chapters 1 and 16, Cartwright (1983) and Ellis (1979), especially pp. 28—9 and footnote 15. Entity realism is strongly supported in Harré (1986) under the name 'referential realism'. This paragraph is in response to a comment by my colleague, Fred Kroon, who also provided other useful suggestions concerning this introduction.
[17] See the useful discussion in Section 1 of Hellman (1983) on which most of the above is based.
[18] For a start see Dummett (1978) the 'Preface' and Chapter 10 entitled 'Realism'.
[19] Putnam (1978), p. 20.
[20] See Putnam (1978), 'Realism and Reason' along with his subsequent views in Putnam (1981) and Putnam (1983).
[21] See Blackburn (1984), Part II, especially Chapter 5, for the 'lifeline he throws across a swirl of philosophical currents' in this area.
[22] See Hellman (1983), Section 3 for these important variants of realism.
[23] Realist views on observation that go well beyond what empiricists might admit are discussed in Shapere (1982) and Nola (1986).
[24] The translation is from Cornford (1935).
[25] 'Reflections on My Critics' in Lakatos and Musgrave (eds.) (1970), pp. 264—5.
[26] Kuhn (1970), pp. 206—7.
[27] Barnes and Edge (1982), pp. 4—5.
[28] Collins (1981), p. 3.
[29] Tarski does, however, view his absolute concept of truth as an instance of a more general relative concept of truth; this does not support philosophical relativism however. See Tarski (1956), p. 199.
[30] Tarski (1956), p. 153.
[31] Newton-Smith (1982), p. 107.
[32] *Ibid.*, pp. 106—8.
[33] Quine (1969), pp. 48—9.
[34] Kuhn (1970b), pp. 266—7.
[35] Hanson (1965), p. 5.
[36] See, for example, Shapere (1982), especially Section IV, Fodor (1984) and Dretske (1969).
[37] See Feyerabend (1978), especially pp. 163—4; see also all of Part I and Section 3 of Part II entitled 'The Spectre of Relativism'.
[38] Hacking (1982), especially pp. 64—5 for his anti-foundational account of styles of reasoning.
[39] See Doppelt (1983).
[40] Barnes and Bloor (1982), p. 34.
[41] *Ibid.*, p. 27.
[42] *Ibid.*, p. 29.
[43] Wittgenstein (1977), Section 105.

[44] *Ibid.*, Section 108.
[45] See *Theaetetus* 169D—171D and Burnyeat (1976) for an excellent unravelling of Plato's various arguments. See Siegel (1986) for two other arguments from Plato.
[46] See the works cited in footnotes 1 to 11; a recent defense of relativism can be found in Margolis (1986).
[47] Barnes and Bloor (1982), pp. 21—2.
[48] *Ibid.*, p. 22.
[49] *Loc. cit.*
[50] *Ibid.*, p. 23.
[51] *Loc. cit.*
[52] Many interesting studies concerning the way people do in fact reason have been carried out by people such as Tversky, Kahnemann, Wason and Johnson-Laird to name a few. For a sample of results in this field see Nisbett and Ross (1980).
[53] This is forcibly argued in Keat and Urry (1975), pp. 204—12 and by Laudan (1981), Sections 3 and 4.
[54] Barnes and Bloor (1982), p. 27.
[55] *Ibid.*, p. 29.
[56] *Ibid.*, p. 41.
[57] For Collins' slightly different line see Collins (1981a) and Collins (1983).
[58] See Collins (1981b) and Collins (1985), footnote 15, pp. 25—6.
[59] Collins (1983), p. 91.
[60] Such dependence claims were discussed above in connection with Barnes and Bloor. They have much in common with the anthropological approach of Mary Douglas; see especially the 'Preface' to Douglas (1975).
[61] The sceptical relativism located here is well expressed in Barnes (1974), p. 154: 'Thus, the epistemological message of the work could be said to be sceptical, or relativist. It is sceptical since it suggests that no arguments will ever be available which could establish a particular epistemology or ontology as ultimately correct. It is relativistic because it suggests that belief systems cannot be objectively ranked in terms of their proximity to reality or their rationality. This is not to say that practical choices between belief systems are at all difficult to make, or that I myself am not clear as to my own preferences. It is merely that the extent to which such preferences can be justified, or made compelling to others, is limited'.

REFERENCES

Barnes, B. (1974) *Scientific Knowledge and Social Theory*, London, Routledge and Kegan Paul.

Barnes, B. and Bloor, D. (1982) 'Relativism, Rationalism and the Sociology of Knowledge' in Hollis and Lukes (eds.) (1982).

Barnes, B. and Edge, D. (eds.) (1982) *Science in Context: Readings in the Sociology of Science*, Milton Keynes: The Open University Press.

Blackburn, S. (1984) *Spreading the Word*, Oxford, Clarendon.

Burnyeat, M. F. (1976) 'Protagoras and Self-Refutation in Plato's Theaetetus', *The Philosophical Review* **85**, 172—95.

Cartwright, N. (1983) *How the Laws of Physics Lie*, Oxford, Clarendon Press.

Collins, H. (1981a) 'What is TRASP?: The Radical Programme as a Methodological Imperative', *Philosophy of Social Science* **11**, 215—24.

Collins, H. (1981b) 'Stages in the Empirical Programme of Relativism', *Social Studies of Science* **11**, 3—10.

Collins, H. (1983) 'An Empirical Relativist Programme in the Sociology of Scientific Knowledge' in *Science Observed: Perspectives on the Social Study of Science* edited by Karin D. Knorr-Cetina and Michael Mulkay; London, Sage Publications.

Collins, H. (1985) *Changing Order: Replication and Induction in Scientific Practice*, London, Sage Publications.

Cornford, F. (1935) *Plato's Theory of Knowledge*, London, Routledge and Kegan Paul.

Davidson, D. (1984) *Inquiries into Truth and Interpretation*, Oxford, Clarendon Press.

Devitt, M. (1984) *Realism and Truth*, Oxford, Blackwell.

Doppelt, G. (1983) 'Relativism and Recent Pragmatic Conceptions of Scientific Rationality' in N. Rescher (ed.) *Scientific Explanation and Understanding: Essays on Reasoning and Rationality in Science*, Lanham, Maryland, University Press of America.

Douglas, M. (1975) *Implicit Meanings*, London, Routledge and Kegan Paul.

Dretske, F. I. (1969) *Seeing and Knowing*, London, Routledge and Kegan Paul.

Dummett, M. (1978) *Truth and Other Enigmas*, London, Duckworth.

Ellis, B. (1979) *Rational Belief Systems*, Oxford, Blackwell.

Feyerabend, P. (1978) *Science in a Free Society*, London, NLB.

Fodor, J. (1984) 'Observation Reconsidered', *Philosophy of Science* **51**, 23—43.

Goodman, N. (1978) *Ways of Worldmaking*, Hassocks, Sussex, Harvester Press.

Hacking, I. (1982) 'Language, Truth and Reason' in Hollis and Lukes (eds.) (1982).

Hacking, I. (1983) *Representing and Intervening*, Cambridge, Cambridge University Press.

Hanson, N. R. (1965) *Patterns of Discovery*, Cambridge, Cambridge University Press.

Harré, R. (1986) *Varieties of Realism*, Oxford, Blackwell.

Hellman, G. (1983) 'Realist Principles', *Philosophy of Science* **50**, 227—49.

Hollis, M. and Lukes, S. (eds.) (1982) *Rationality and Relativism*, Oxford, Blackwell.

Husserl, E. (1970) *Logical Investigations*, volume 1, translated by J. N. Findlay, London, Routledge and Kegan Paul.

Keat, R. and Urry, J. (1975) *Social Theory as Science*, London, Routledge and Kegan Paul.

Kuhn, T. S. (1970a) *The Structure of Scientific Revolutions* (Second Edition). Chicago, The University of Chicago Press.

Kuhn, T. S. (1970b) 'Reflections on My Critics' in Lakatos and Musgrave (1970).

Lakatos, I. and Musgrave A. (eds.) (1970) *Criticism and The Growth of Knowledge*, Cambridge, Cambridge University Press.

Laudan, L. (1981) 'The Pseudo-Science of Science?', *Philosophy of Social Science* **11**, 173—98.

Lenin, V. I. (1908). *Materialism and Empirio-Criticism: Critical Comments on a Reactionary Philosophy*, English translation 1970, Moscow, Progress Publishers.

Margolis, J. (1986) *Pragmatism Without Foundations: Reconciling Realism and Relativism*, Oxford, Blackwell.

Meiland, J. W. (1977) 'Concepts of Relative Truth' *The Monist* **60**, 568—82.

Meiland, J. W. (1979) 'Is Protagorean Relativism Self-Refuting?', *Grazer Philosophische Studien* **9**, 51—68.

Meiland, J. W. and Krausz, M. (eds.) (1982) *Relativism: Cognitive and Moral*, Notre Dame, University of Notre Dame Press.

Merrill, G. M. (1980) 'Three Forms of Realism' *American Philosophical Quarterly* **17**, 229—35.

Newton-Smith, W. H. (1982) 'Relativism and the Possibility of Interpretation', in Hollis and Lukes (eds.) (1982).

Nisbett, R. and Ross, L. (eds.) (1980) *Human Inference: Strategies and Shortcomings of Social Judgement*, Englewood Cliffs, N.J., Prentice-Hall.

Nola, R. (1986) 'Observation and Growth in Scientific Knowledge', *PSA 1986*, Volume One, edited by A. Fine and P. Machamer, Philosophy of Science Association, East Lansing, 245—57.

Passmore, J. (1961) *Philosophical Reasoning*, London, Duckworth.

Putnam, H. (1978) *Meaning and the Moral Sciences*, London, Routledge and Kegan Paul.

Putnam, H. (1981) *Reason, Truth and History*, Cambridge, Cambridge University Press.

Putnam, H. (1983) *Realism and Reason: Philosophical Papers Volume 3*, Cambridge, Cambridge University Press.

Quine, W. V. (1969) *Ontological Relativity and Other Essays*, New York, Columbia University Press.

Rorty, R. (1982) *Consequences of Pragmatism: Essays 1972—80*, Brighton, Harvester.

Shapere, D. (1982) 'The Concept of Observation in Science and Philosophy', *Philosophy of Science* **49**, 485—525.

Siegel, H. (1986) 'Relativism, Truth and Incoherence', *Synthese* **68**, 225—59.

Sober, E. (1982) 'Realism and Independence', *Nous* **16**, 369—85.

Tarski, A. (1956) *Logic, Semantics, Metamathematics*, Oxford, Clarendon Press.

Wittgenstein, L. (1977) *On Certainty*, Oxford, Blackwell.

DAVID PAPINEAU

DOES THE SOCIOLOGY OF SCIENCE DISCREDIT SCIENCE?

1. THE NEW SOCIOLOGY OF SCIENCE

Why should we believe the findings of science? The traditional answer is that there are good *arguments* for those findings. Scientists start from the evidence of their senses, and then proceed by valid inferences to their conclusions. The scientific community is the rational community *par excellence*, and the findings of science are simply the conclusions forced on scientists by the dictates of reason.

This traditional view of science's epistemological authority is complemented by traditional sociology of science. Traditional sociology of science analyses scientific institutions. It looks at the norms governing such things as scientific publications, scientific education, priority disputes, fraud accusations, etc. But within traditional sociology of science there is no sociological analysis of scientific reasoning and its products themselves. The real engine of science — the processes that fix the contents of scientific theories, rather than their mere existence and subject matter — is simply taken for granted. The engine of science is simply the power of pure reason, and as such is not a fit subject for sociological analysis.

In the last decade or so a "new sociology of science" has emerged, distinguished by its readiness to investigate the processes which actually influence the *content* of scientific theories. And a central lesson of the new sociology of science seems to be that the actual behaviour of science bears little relation to the traditional rational picture. The factors influencing real scientists in their theoretical decisions do not look like good arguments at all.

We can usefully distinguish a macro-sociological and a micro-sociological school within the new sociology of science. The former school, whose stronghold is the Science Studies Unit at the University of Edinburgh, favours explanations of scientific theories in terms of their ideological significance, or their consonance with other features of the general social context. Thus, for example, the Darwinian theory of natural selection has been associated with certain social doctrines

Robert Nola (ed.), Relativism and Realism in Science, 37—57.

prevalent in Victorian England. Pasteur's success in 'disproving' the thesis of spontaneous generation has been attributed to the association of this thesis with such atheistic views as materialism and evolution, in the context of the conservative and religiously orthodox Second Empire. Or again, to take a rather different kind of case, David Bloor has suggested that the argumentative strategies adopted by mathematicians, and in particular the nature of their reactions to counterexamples, reflect the social structure of their mathematical community, and in particular the nature of its internal and external social divisions. [Cf., respectively, Barnes and Shapin (1979), Farley and Geison (1974), Bloor (1978).]

The micro-sociological branch of the new sociology of science, on the other hand, has affinities with the general ethnomethodological movement in sociology. As such it is by and large uninterested in large-scale social influences on scientific developments. Instead it studies the micro-interactions which lie behind the achievement of scientific consensus. It emphasizes the negotiations, the trial and error procedures, and the tacit understandings which precede any agreement on scientific findings. A significant factor in such negotiations is often the concern of individual scientists, or small groups, to gain credit for "establishing facts". Such establishing of facts is viewed as a matter of persuasion and argument. What kinds of experiments are conducted, what kind of controls are required, how the results are presented, where the papers are published, etc., are all explained in terms of the need of scientists to win acceptance for their "facts" from their colleagues and opponents. Often there will be significant periods of vagueness and negotiation during which the different parties will compete and manoeuvre, and when the fate of a putative fact will hang in the balance. Success in winning acceptance for facts will depend in part on the contextually variable norms of the particular scientific community, and in part on the *resources* the scientist can command: most obviously, resources in terms of money and equipment, but also, less tangibly yet more fundamentally, resources in terms of "scientific credibility", which gets built up, like capital, during a career, and which can then be re-invested in pursuit of further "facts". [Perhaps the best-known example of such micro-sociology of science is Latour and Woolgar's *Laboratory Life* (1979). See also H. Collins (1985), M. Lynch (1985).]

My question is this: does the new sociology of science discredit science? Should those of us who have hitherto been inclined to believe the findings of science be worried by these new descriptions of scientific activity?

My answer will be that it depends. It depends on one's attitude to epistemology. Those with certain orthodox, "Cartesian" assumptions about the nature of epistemology are indeed right to be worried by the new sociology. But I shall show that there is an alternative, *naturalized* approach to epistemology which makes it possible to accommodate the new sociology without rejecting science.

One point before proceeding. As these last remarks will have suggested, I shall not be directly concerned with the factual standing of the new sociology. Perhaps some of its claims are false. No doubt many of its claims (especially in the macro-sociological mode) are speculations that demand further empirical research. But since my concern is really with the hypothetical question, "What *would* follow *if* the new sociology were true?", I am happy to take it as read that there is real factual substance in the new sociology.

2. TWO KINDS OF EPISTEMOLOGY

Epistemology is a practical business. It tells us what to do in order to ensure we have the right beliefs. More specifically, an epistemological *theory* specifies a certain kind of preferred *technique* for acquiring beliefs, and then makes the normative *recommendation* that concerned believers should acquire all their beliefs from that technique. (When an actual belief derives from the preferred technique, we say it is *justified*.)

The *Cartesian* theory of epistemology recommends that we should get our beliefs from *good arguments*. We should assent only to those beliefs that have been generated by logically valid steps from secure premises. An actual belief is justified just in case in issues from such an argument.

This Cartesian conception of epistemology lies behind most of the Western philosophical tradition of the last three hundred years. Even today nearly everybody takes it for granted. It was implicit in the opening remarks of this paper about the reasons for believing the findings of science. But there is an alternative way of thinking about epistemological matters. There is another epistemological theory, the naturalized theory, which gives an equally powerful account of how to get the right beliefs.

I shall explain this naturalized alternative in a moment. But first it is worth noting how the Cartesian theory goes hand in hand with the conception of the conscious mind as transparent to itself. We are to assent only to those beliefs that follow validly from secure premises. But how are we to select those secure premises? And how, for that matter, can we ensure that the steps leading from them to the conclusion are indeed valid? By further arguments? In order to avoid the regresses threatening here we need some beliefs whose truth is ensured without argument, and some logical steps whose validity is immediately apparent. From Descartes onwards these have been provided by the conscious mind's supposed privileged awareness of its own ideas and of the relations between those ideas.

Now for the naturalized theory. On this theory, the right technique for acquiring beliefs is simply to be a *reliable belief-former*, that is, to have belief-forming processes that generally produce true beliefs. Concerned believers should try to ensure that all their beliefs come from belief-forming processes that are reliable in this sense. An actual belief is justified just in case it issues from a reliable process. [Cf. Goldman (1976), (1979).]

The difference between the naturalized and the Cartesian theory is best brought out by considering how the naturalized theory deals with perception and memory. According to the naturalized theory, perceptual beliefs and memories are justified if they come from reliable processes, and are unjustified if they come from unreliable proceses. So a good example of an unjustified belief would be little Johnny's naive belief that the pitch of the train's whistle changes as the train passed by: for the process behind this belief is unreliable, in that it characteristically generates false beliefs. Similarly, people who naively succumb to *déjà vu* experiences, and take themselves genuinely to be seeing things for a second time, will have unjustified beliefs. For, again, the processes behind their beliefs are unreliable. Of course not all the perceptual and memory processes embodied by human beings are unreliable in this way. In most cases our perceptual and memory beliefs come from reliable process and then, the naturalized theory wants to say, those beliefs are justified.

So the naturalized notion of justification doesn't necessarily involve conscious argument. On the naturalized approach, the difference between justified and unjustified perceptions and memories *can't* be due to the former having superior argumentative backing: for, after all, the distinguishing characteristic of perception and memory is that from the perspective of consciousness they are *non*-inferential. Our perceptual and memory processes deliver beliefs into consciousness, but the processes leading up to those beliefs lie outside consciousness.

Not that the naturalized approach rejects argument altogether. For argument is itself a belief-forming process. Given existing beliefs as premises, argument is a process that moves us to new beliefs as conclusions. So, as part of the general recommendation that our belief-forming processes should be reliable, the naturalized approach recommends that our argumentative habits should be reliable too. (Though of course here the appropriate notion of reliability is conditional, rather than categorical: the conclusions should be true *if* the premises are.)

So the naturalized approach certainly allows that a belief's justification can be due to its coming from a good argument: if the premises of an argument are themselves reliably generated then the conclusion of that argument is justified just in case the argumentative habit in question is conditionally reliable for truth. But note that the naturalized attitude to argument is here quite different from the Cartesian attitude. It is not because argument is a conscious phenomenon, transparent to the mind, that it is good for generating beliefs. The point is simply that certain kinds of argumentative habits reliably generate true conclusions given true premises. Consciousness doesn't come into it. From the naturalized perspective, if a completely non-conscious being embodied reliable inferential processes, then those processes would give that being justified beliefs (or "beliefs", if you prefer).

I take it to be the overriding virtue of the naturalized approach that it doesn't need consciousness's supposed power of incorrigible self-awareness. It allows us to replace the picture of autonomous individual subjects, each locked into the mental arena of their own epistemological responsibilities, with a picture in which human thinkers are simply normal beings interacting with the rest of the natural world.

A corollary is that the naturalized approach allows us to get away from the idea that epistemology is necessarily about techniques which a single individual can use to acquire beliefs. So far in this section I have not questioned this assumption, and have concentrated on the attitude of naturalized epistemology to processes embodied in single individuals, like perception, memory and argument. But there is no need to go on thinking of epistemology in this way once we reject the idea that all epistemological techniques must operate within conscious minds. If we switch to the naturalized notion that justified beliefs require reliable processes, processes whose operation may well lie outside consciousness, we can perfectly well think of beliefs as justified because they come from reliable *social* processes, processes which involve whole sets of people and the interactions between them, and not just from processes operating in individuals. This will be important in Sections **5** and **6** below.

3. MORE ABOUT NATURALIZED EPISTEMOLOGY

A full investigation of the strengths and weaknesses of the naturalized approach is out of the question here. But given the relative unfamiliarity of the naturalized approach, it will be worth saying a bit more about it, and in particular about its relevance to the normative question of what we *ought* to believe.

Many philosophers find it difficult to see how the naturalized theory can possibly be an answer to this practical question. If we try to follow the naturalized theory's normative recommendation that we ought to acquire all our beliefs from reliable processes, we seem to face an insuperable dilemma. The demand for naturalized justification can be read either "externally" or "internally". If we think of it entirely "externally", with justification depending on the reliability of processes some of which lie outside the domain of consciousness, then how can such "justification" be of any significance to a concerned believer? Such a believer won't necessarily know whether or not the processes involved are reliable. So how can the fact of their reliability or unreliability possibly influence the concerned believer in deciding what to believe?

Alternatively, we can read the naturalized approach "internally", as recommending that for any given belief we need to have a further belief to the effect that the original belief was reliably produced. But then we face a particularly vicious version of the familiar regress. (What about the further belief about reliability? Do we need to believe that *that* was reliably produced? Etc.)

But this dilemma begs the question. It seems insuperable only because the grip of the Cartesian approach is so strong. It takes it for granted that what the concerned believer needs is arguments. Thus: if the relevant facts are unknown, they cannot enter into the concerned believer's *arguments*; if we do try to get them into the concerned believer's *arguments*, there is an obvious regress.

The question, however, is precisely whether concerned believers *should* seek out arguments, or whether they should simply aim to be reliable belief-formers. Naturalized epistemology makes the latter recommendation. But this should not be understood as a recommendation as to how better to conduct arguments. That is the very point at issue.

It might still be quite unclear how the naturalized theory can have any normative import. But once we free ourselves from the idea that argument is the only possible epistemological technique, nothing could be simpler. You want to be a reliable belief-former? Well then, you'd better do what has to be done to bring this about. You'd better investigate what belief-forming processes you already embody, and you'd better consider what alternative such habits you might adopt. And in particular you'd better investigate which of those habits, actual and possible, are reliable for generating truths. And having done all that, you should take steps to rid yourself of any bad, unreliable habits you already have, and take steps to instil any good, reliable ones that are open to you.

In effect, the idea is to think of yourself as a system for generating true beliefs. You want to be as reliable a system as possible. So you consider ways of redesigning the system, and you implement those that promise an improvement. [Goldman makes some suggestions along these lines in his (1978), (1980), and (1985). But it seems to me that he fails fully to free himself from Cartesian presuppositions: see in particular p. 30 of his (1980), and p. 40 of his (1985).]

Further questions obviously remain. Maybe we can sensibly *aim* to be reliable believers by reflecting on our belief-forming processes, and practising those we judge reliable. But will this strategy necessarily *succeed* in making us reliable belief-formers? In particular, isn't there an obvious difficulty? Our judgements as to which belief-forming processes are reliable will inevitably depend on our existing beliefs about the world and how we fit into it. So won't the naturalized recommendation at best get us into the state of our belief-forming processes *seeming* reliable (in the light of our existing beliefs), not the desired state of their being reliable?

Here the story gets complicated. In favour of the naturalized approach, note that it is by no means automatic that any given belief-forming process will vindicate itself as reliable. Somebody who habitually succumbs to *déjà vu* is quite capable of investigating this disposition of mind, as a psychologist might, and of concluding that in general *déjà vu* "memories" do not have the facts that would make them true among their antecedents. More generally, the beliefs that *issue* from a given belief-forming process (such as: I've been here before) are not as a rule the beliefs which will *justify* that process as reliable (for this you would need something like: in general, whenever somebody has a *déjà vu* feeling, this is a result of that person having previously experienced the relevant scene).

So the naturalized epistemologist can certainly argue that the naturalized recommendation has some bite: it won't automatically leave everybody with whichever belief-forming processes they might naively have happened to start with. But still, even if the naturalized strategy leads you somewhere, this won't necessarily mean that it will lead you to reliable methods (as opposed to reliable-seeming methods). Might not different communities, with different intellectual starting-points, each apply the naturalized strategy, and yet get led off in different directions *ad infinitum*? Advocates of naturalized epistemologist will want to deny that this is in fact possible. They will urge that the world itself will act as a rudder steering different communities towards the truth (provided, of course, that those communities play their part by taking pains to ensure that their belief-forming dispositions are as well-adjusted to the world as possible).

Clearly there are questions of argumentative onus here. This last thought scarcely *proves* that the naturalized recommendation will lead to truth. But naturalized epistemologists will deny that they have any obligation to show that their method will *inevitably* be successful. Their claim is simply that it will, as it happens, succeed, and they will point out that there is no obvious reason to expect the world to frustrate such success. They will say that the desire for proof we feel here is itself a hangover from the Cartesian way of thinking: if epistemological justification did depend on Cartesian arguments transparent to the mind, then it would indeed follow that if you were justified you could prove that you were; but if justification depends only on being well-adjusted to the world, then there is no reason why being able to prove that you are justified should be a necessary condition of being justified.

Naturalized epistemology faces other problems. In giving examples of the evaluation of belief-forming methods I have concentrated on non-inferential processes like perception and memory. In such cases it is easy enough to see how reliability-evaluations are supposed to work, even if you have doubts about their further significance. But what about inferential methods like deduction and induction? In particular, what about the kind of theoretical inferences that move scientists from their experimental findings to their theoretical conclusions? Here matters are far less clear, and showing that our inferential habits are well-designed to lead us from truths to truths is a substantial research programme for the naturalized epistemologist. (Note, however, that it's not the impossible programme of showing somebody who has no such habits that they ought to adopt some; rather it's the task of explaining to ourselves, in the light of our beliefs – including our inferred beliefs – about the general structure of the world, why it is that our inferential practices work.)

This is not the place to try to convince you of the superiority of the naturalized approach. But I hope I have done enough to show that it is a serious competitor to Cartesianism. Of course there are problems facing naturalism. But that is scarcely a reason for continuing to view all epistemological issues through Cartesian spectacles. After all, three hundred years of philosophical toil notwithstanding, there are plenty of familiar problems facing Cartesianism too. We shall be reminded of some of these in the next section. [For further defence of the naturalized approach to epistemology, see Papineau (1987).]

4. CARTESIANISM AND THE NEW SOCIOLOGY

In this Section I want to explain why Cartesians will take the new sociology of science to discredit science. The answer to this question might seem obvious. Don't Cartesians take the worth of science to depend on the supposed purity of the scientific mind, on the supposed fact that the scientific consciousness is swayed by reasoned argument alone? And doesn't the new sociology of science shows this supposed fact is not a fact at all?

However, there is a possible response open to Cartesians who want to defend science against the new sociology. Maybe, they can allow, scientists are on occasion influenced by ideological prejudice, or by personal ambition, or by other unscientific motives. But, they can then insist, it by no means follows that the views those scientists espouse are to be rejected. For there may still *be* good arguments for those views, in the sense that those views may still be ones that an ideally rational person acquainted with the evidence *would* uphold. That is, Cartesian defenders of science can appeal to the distinction between the normative question of which views ought to be accepted, and the genetic question of what actually led certain particular people to adopt those views.

An immediate difficulty facing this response is that it gives no grounds for a *general* faith in scientific findings. Maybe specific scientific theories can be independently vindicated by appeal to abstract standards of reason. But if the causes which actually operate in persuading scientists to accept theories are quite independent of the abstract reasons that can so normatively vindicate those theories, won't it be a freak if the findings of science are generally belief-worthy?

At this point a Cartesian needs to appeal to some kind of "invisible hand", some kind of unintended social mechanism which will ensure that in general the personal motives of scientists will always cancel out and leave us with belief-worthy theories. Now, perhaps a reasonable case can be made for such an invisible hand: perhaps there is a story about technological success and funding support, say, which will show that only the good theories survive in science. Indeed I myself shall argue for a somewhat different kind of invisible hand in Section **6** below. But let me leave this question up in the air at this point. For I now want to show that, whatever we think about invisible hands, there is a rather deeper reason why Cartesians have to resist the new sociology if they are to defend science.

I have been speaking of 'good arguments' and the 'dictates of reason'. But what are these dictates of reason? What authenticates the abstract normative principles with which the Cartesian hopes to vindicate scientific theories?

When we see how this question has been answered in recent philosophy of science, we will see why the new sociology of science raises a quite particular and intense difficulty for the Cartesian friends of science. But first we need a bit of general stage-setting. Note first that there is a strong tendency for Cartesians to reject *realism*. By realism I mean the view that success in judgement consists in correspondence to an independent reality. Cartesians who are realists face an uphill task. They have to show from first principles that the standards of reason which (a) recommend themselves to conscious human minds are (b) guaranteed to produce beliefs that correspond to an independent reality. It is not at all clear how to show this. So the natural move is to reject (b) and embrace *anti-realism*.

By anti-realism I mean the position that makes *reason* prior to *truth*.

Where the realist starts off with the notion of truth as correspondence (and so incurs an obligation to show that reasons are a good guide to truth), the anti-realist takes reason to stand on its own feet. The anti-realist takes success in judgement to consist directly in beliefs being backed by reasons, in their having the right provenance, rather than in terms of their correspondence to reality. And so for the anti-realist there is no question of showing that reasons are a good guide to truth. 'Truth' and 'reality', if they are used at all, are simply epithets attached to the picture of the world that reason leads us to. 'Truth' and 'reality' become by-products of rationality, rather than its aim.

Of course, not all Cartesians have been anti-realists. Descartes wasn't, for one. Nor was Locke. But Descartes and Locke could appeal to a God-given natural light to link up reason and realist truth. This had ceased to be a philosophical option by the middle of the eighteenth century. And since then the Cartesian tradition has unquestionably been overwhelmingly anti-realist.

So I have it that Cartesian epistemology leads to anti-realism. What now of the question raised above, "How can Cartesians authenticate the standards of reason?" It is all very well being told that reasons should be introspectively available, that they should automatically recommend themselves to consciousness. But can consciousness be sure that it has hold of the right reasons? Mightn't we be mistaken in assuming that inductive support, say, or simplicity, were good reasons for believing some scientific claim?

At this point the standard anti-realist strategy is to reject the problem. Anti-realists will argue that doubts as to whether intuitively good reasons really are good reasons only arise if we succumb to the realist misconception that the job of reason is to produce beliefs that correspond to reality. Perhaps, the anti-realist will concede, there are no higher meta-arguments which *prove* that inductive support, or simplicity, are guaranteed to produce true beliefs. But why demand such meta-justifications? Given anti-realism, there is no need to ask whether human standards of rationality are well-suited to generating true beliefs: a belief's being true just *is* a matter of its being generated by human rationality.

But there remain obvious worries about the 'human standards of rationality' being invoked here. We seem to be presupposing that standards of rationality are universal. But don't different people have different standards of rationality? Some anti-realists take a short line here too. They simply argue that, despite any appearances to the contrary, alternative rationalities are not possible: the very idea of a community with radically different standards for evaluating beliefs makes no sense. Thus, for instance, some interpreters of the later Wittgenstein take the moral of his stories about people who measure piles of wood by cross-section rather than volume, etc., to be that such apparent alternatives are incoherent, that these alternative possibilities of thought are not real possibilities after all. [Cf. Lear (1982).] And the later Putnam has argued similarly that we have no alternative but to deem people who diverge from our basic standards of rationality to be crazy, to be incapable of human thought at all [Putnam (1981), Ch. 6].

However this blunt denial of any possibility of alternative rationalities will seem too quick to anybody working in the philosophy of science, to anybody concerned specifically with the rationality of scientific theory choice. For it is manifestly *possible* for people to suppose that modern astrology, or creationism, are supported by reason. And it would be tendentious at best to insist that such people must be crazy.

A natural ploy at this point is for the anti-realist to shrink the sample of humans whose thought is taken to manifest rationality. The anti-realist can, so to speak, reduce the data base against which hypotheses about the right rationality are to be tested. Not all human thinkers are *per se* rational. Rather it is the mature, serious thinkers, those with the time and resources and inclination to investigate nature critically, to whom we should look to identify the real standards of rationality. In short, it is the intellectual activity of *scientists* which serves to define the right reasons for belief.

And this, in effect, is the view of scientific rationality adopted by many contemporary philosophers of science: proposed methodologies, proposed sets of standards for scientific theory choice, are to be evaluated by comparing them with the actual intellectual practice of past scientists. The way to identify the right methodology for doing science is to think of proposed philosophies of science as themselves *theories*, which can then be tested against the data provided by the history of science. The right methodology is the one that best fits the history of science. [See in particular Lakatos (1978); Laudan (1977).]

Taken on its own this approach can seem puzzling. How can we hope to get normative conclusions about the right way to do science from factual premises about the way certain past people have reasoned on past occasions? But this only remains a puzzle if you think of human intellectual practice as answering to some higher standard (such as, for instance, the need to generate conclusions that correspond to an independent reality). If there isn't anything more to the 'right' standards of rationality than simply the standards that come naturally to mature human thinkers, then how else should we identify those standards except by familiarizing ourselves with the habits of thought of the central figures in our scientific tradition? This way of adjudicating between philosophies of science may seem odd when taken out of context. But it makes perfect sense when seen against the background of Cartesian anti-realism.

So for Cartesian anti-realists, especially for those working in recent philosophy of science, rationality is by definition the way that scientists think. But now the intense threat posed to Cartesians by the new sociology of science should be clear. If you distil rationality from the history of scientific practice, then you are no longer in any position to argue that although scientists are often moved by ulterior motives, scientific theories themselves are in the abstract rationally supportable. For the things that move scientists are *per se* what's rational. And so, if scientists are no better than the rest of us, if their thoughts too are characteristically swayed by ideological prejudice and social context and self-interest, then it seems to follow that there is no such thing as rationality after all, beyond a degree zero a-rationality which leaves all beliefs, however ill-founded, on a par. For those who aim to found rationality on scientific practice, the new sociology of science discredits not just the motives of scientists, but the very idea of rationality itself.

Let me sum up the reflections of this section. I do not want to suggest that all Cartesians necessarily have the rationale I have outlined for resisting the new sociology. Just as there are some Cartesians who are not anti-realists (Descartes, Locke), so there are some anti-realists (Wittgenstein, Putnam) who do not attempt to distil scientific rationality out of the history of science. But, even so, there are a significant number of contemporary philosophers who have been led to attempt this, and I hope I have made it clear why these philosophers at least cannot allow space to the new sociology. [In this connection, it is worth observing that those philosophers whose history-based notion of rationality is most explicit, and whose anti-realism is correspondingly closest to the surface, are also those who are most insistent in resisting the inroads of sociology. Both Lakatos (1978) and Laudan (1977) are quite adamant that sociology has no place whatsoever in the analysis of proper science.]

5. NATURALISM AND THE NEW SOCIOLOGY

I now want to consider what a naturalized epistemologist ought to say about the new sociology of science.

Contrast the situation of naturalism with that of Cartesianism. In the last section I argued that there were two routes by which the new sociology threatened Cartesian defenders of science. In the first instance, it raised immediate questions as to whether scientific findings were generally the result of good arguments. And then, in response to the thought that an invisible hand might ensure that in the abstract scientific findings were always rationally support*able*, even if the scientists had not so supported them, the new sociology threatened the very idea of rational support itself.

Neither of these discrediting implications go through if we adopt the naturalized perspective. Take the second threat first. This does not arise for naturalism, for there is nothing to draw naturalists to the anti-realist idea that rationality is nothing but those ways of thought natural to humans, and so, *a fortiori*, there is nothing to draw the naturalist to the idea of distilling standards of rationality from the history of science. For naturalists, a belief is rational just in case it is produced by a belief-forming process that is reliable for truth. Since naturalism here rejects the whole idea that justification demands arguments whose cogency is transparent to consciousness, naturalists by-pass the problem, faced by Cartesians, of showing from first principles that conscious human rationality is well-suited to generating beliefs that correspond to reality. And so there is no motive whatsoever for naturalists to become anti-realists. To put it at its simplest, naturalists build the idea of producing beliefs that correspond to reality into the very idea of rationality. And so, whatever other difficulties this approach might raise, they are not going to be difficulties which force naturalists to a truth-independent notion of rationality.

Consider now the first, more immediate, threat raised by the new sociology: it shows that in general scientific findings are not the results of good arguments. But why should this worry a naturalized friend of science? Once more, the naturalized epistemologist does not think of the belief-worthiness of science as depending on there being good *arguments* for the findings of science: the requirement is only that those findings should generally be produced by reliable methods. And it is by no means immediately obvious that the new sociology succeeds in casting doubt on *this* requirement.

This last observation is the central point of this paper, and I want to develop it in some detail. But first let me make one thing clear. I am not here producing arguments to show, nor am I presupposing, that the findings of science are in fact generally true. I do, as it happens, believe this. And I also, as it happens, believe that the naturalized perspective allows us to see, in general terms, how it is that we are entitled to this belief. But that is not my present project (though the remarks introducing naturalized epistemology in Sections **2** and **3** above give some indication of how this larger project might be carried out). All I want to establish here is that a certain *negative* argument *against* believing in science does not carry the weight it seems to: that is, I simply want to show that, given naturalized epistemology, the new sociology of science should not disturb any faith in science you might independently happen to have.

My central claim, then, is that the new sociology of science does nothing to show that scientific practice is not generally reliable for generating true theories. It may well show that scientists are often swayed by prejudice, ambition and other ulterior motives. It may well show that the internal mental motivations of scientists are no different from those of the general public. But it by no means follows that the overall structure of scientific practice is not reliable for truth.

Let me elaborate this point in connection with that central plank of the new sociology, the "symmetry thesis" that true beliefs should receive just the same kind of explanation as false ones. [Cf. Bloor (1976), Ch. 1.] In a sense the naturalized friends of science can happily admit this thesis. They can accept that all scientific beliefs are produced by natural processes, and that there is no need for any special category of conscious reasons to play a special part in the generation of scientific beliefs. But they do need to insist on one point. The natural processes generating scientific beliefs can not, if they are to be reliable for truth, include *only* social factors, if this is understood to exclude the influence of the truth conditions of those beliefs, the facts that the beliefs in question are supposed to be about. Thus, to take an analogous case, while the reliability of our long-term memory does not demand that we derive our memories from some kind of Cartesian cogitation, it clearly does require that amongst the causal inputs to the processes generating our memories should be the facts those memories are about.

So there is a minimal demand to be made by the naturalized friend of science: the explanation of creditable beliefs needs to differ from those of beliefs in general at least to the extent of allowing that amongst the causes of those beliefs are the truth conditions of those beliefs. But this now is why the new sociology fails to discredit science: it gives us no reason to suppose that this minimal demand is not satisfied. That the relevant natural facts play *no* part in the causation of scientific theories is an extremely strong claim, which is by no means established by the empirical findings of the new sociologists.

While they are not always explicit on the matter, many new sociologists seem to *think* that their work discredits the findings of science. But I would say that, if they do, this is because they are taking Cartesianism for granted. For a Cartesian, as we have seen, the fact that social influences play some part — a large part — in influencing the mind of the scientist, is indeed enough to discredit science. But something much stronger — that social factors play the *only* part — is needed to persuade the naturalized epistemologist against science.

Let me consider the two branches of the new sociology in turn. According to the macro-sociologists of science, scientists believe theories that fit their ideological position, or are otherwise consonant with the large-scale social context. If it were in general true that such social influences were *sufficient* for the acceptance of scientific theories, then we (I now abandon any further pretence of impartiality and identify myself with the naturalized epistemologist) should indeed be suspicious of science, for any conformity between scientific theories and the facts they were about would be purely happenstantial. But it does not matter that much if such social factors are *necessary* for the acceptance of theories. Thus, to take a familiar example, we could allow that a climate of high competitive capitalism was necessary for the acceptance of natural selection theory (and I mean the *acceptance*, not just the 'discovery', the thinking up of the theory — we can allow that natural selection would not have been accepted even if it had been proposed, as it arguably was by Buffon, in the *ancien regime*) and still hold that the social factors weren't *sufficient* for acceptance: that, in short, the truth of the theory was necessary as well. Provided the relevant natural facts are necessary for acceptance along with the social ones, there is no reason to doubt that scientific practice as a whole is reliable for truth.

(We should of course allow that sometimes, as in the Lysenko episode, the acceptance of theories is forced by social facts alone. But that such things can on occasion happen, in special circumstances, does not show that *scientific* institutions will in general allow such purely social determination of theories.)

Let me now turn to the micro-sociologists of science. It is interesting to note that from a naturalized perspective many of the descriptions micro-sociologists give of scientific practice are extremely reassuring about the reliability of science. No doubt it is true that scientists are often primarily motivated to attach their names to facts, and to build up the scientific credibility that will enable them to do this. But consider how they go about doing this. A prime concern, if they are to persuade others, is to ensure that their opponents won't be able to pick holes in their published claims. So scientists take care that their experiments are repeatable. They try to design experiments so that the move from the observed results to the desired theoretical conclusions depends on as few disputable auxiliary hypotheses as possible. They try to devise a variety of experiments, with a suitable range of controls, so as to leave their opponents no room for manoevre. And so forth.

To the naturalized epistemologist this is likely to seem just what is needed to ensure that scientific acceptance is reliable for truth. (And even the periods of "negotiation", when the fates of putative new facts hang in the balance, can be interpreted as periods of healthy agnosticism, during which further analysis and experimentation is needed to decide whether the new claims should indeed be entered in the archives.)

6. THE GAME OF SCIENCE

Consider this argument. "There is no compelling transcendent logic which forces scientists to their conclusions. That certain facts get written into the archives depends on socially contingent processes involving power, negotiation, and argumentative work. So scientific findings are merely social constructions, with no further epistemological authority."

At one level the appropriate response is obvious. It is only within the Cartesian perspective that justification requires arguments whose logic is compelling to consciousness (or at least to the consciousness of mature, scientific thinkers). Naturalism, by contrast, allows that justified beliefs can come from whatever processes you like, however contingent or arbitrary they might seem from the point of view of conscious logic, as long as those processes are reliable for truth.

But the underdetermination of scientific practice by simple logic does point to a more substantial difficulty. If sound scientific habits of thought are not logically inescapable, is there not, then, a pressing need to explain why scientists do adopt their peculiar habits of thought? And surely at this level the only possible explanatory factors will be social ones.

And if so, won't this in itself discredit science? Put it like this. Maybe, say, the particular community of Californian endocrinologists studied by Latour and Woolgar do indeed embody habits of thought that are reliable for truth. But if they are sociologically caused to have those habits of thought, isn't there a sense in which they are just *lucky*? For all *they* have contributed to the enterprise, for all they know and care about, they could as well have been caused to adopt *un*reliable habits of thought.

And, more generally, if sound scientific practice is just a matter of luck, what is to ensure that scientists in general, as opposed to such particular communities as Californian endocrinologists, will have reliable habits of thought? In Section **2** I touched on the possibility of Cartesians evading their initial difficulties by positing an "invisible hand". But now I seem to be committed to an invisible hand myself, and indeed one that is required to do rather more work: for where the Cartesians just wanted a mechanism to select out rationally supportable *beliefs*, I seem to need a mechanism which will ensure that the belief-forming *methods* of scientists in general are reliable producers of truths.

It might seem that naturalized epistemology can once more dig in its heels and deny the difficulty. What does it matter *why* scientists have reliable habits of thought, provided that they *do* have them? Maybe the Californian endocrinologists *are* just lucky. Maybe it *is* just a matter of luck that scientists in general think in ways that produce truths. But why bemoan such luck, if it indeed obtains?

There is indeed a sense in which the fact of reliability is more important than its explanation. But, still, it would be a puzzling freak if scientists in general adopted reliable habits of thought, and yet there was no general explanation of why they did so. And, in fact, if that were the case, any claim to the effect that scientists are in fact generally reliable would lose all plausibility: piecemeal demonstrations that particular scientific communities at particular times happened to have reliable processes would not prove anything, if we did not think there was some general explanation for this.

But fortunately there is an explanation. Namely, that scientists are their own naturalized epistemologists. Scientists are themselves perfectly capable of reflecting on what is required for the reliable production of truths. They are aware that unrepeatable experiments, or inferences based on *ad hoc* hypotheses without independent support, are unreliable guides to the facts. More significantly, scientists working in any particular speciality will have detailed ideas about the alternative hypotheses that need to be taken into account, the kinds of experimental controls that are required, the experimental techniques that are called for, etc., if truths about their specific subject matter are to be got at. And they will use this expertise to dismiss and discredit claims to knowledge that are based on unreliable methods.

In effect I am here appealing to a point made back in Sections **3** and **4**: it is perfectly possible for human beings to reflect on whether the belief-forming practices they engage in are reliable for producing truths, and to reconstruct themselves and their communities in the light of such reflections. As we saw then, it is an open question, which raises delicate issues of argumentative onus, as to whether such self-adjustment in pursuit of reliability will succeed in producing beliefs that actually are reliable (as opposed to ones which merely seem reliable). But we do not have to resolve this issue here. For the present objection is the specific one that it would at best be an accident if scientific practice was generally reliable (and that therefore we can not suppose that it is). And it is sufficient answer to this objection to point out that, whether or not the self-adjusting attempts of scientists to be reliable are *guaranteed* to succeed, it will scarcely be an *accident* if reliability is in fact the upshot of such efforts.

This last suggestion might seem to be going back on my factual concessions to the new sociology. Aren't I now denying that scientists are swayed by ulterior motives, and instead presenting them as impartial seekers after truth? But the conflict here is only apparent. Take the analogy of a sport like tennis, or, perhaps better, a gambling game like poker. Very few tennis players or poker players are primarily motivated by the desire to conform to the rules of their game. They are after money, or glory, or self-esteem, or perhaps just the satisfaction of winning. And they will use such resources as they have at their command to achieve their ends: their psychological or physical advantage, or their superior skill, or simply their bigger bank-roll. Indeed many will be prepared to break or bend the rules, if they can get away with it, in order to succeed. But, still, there is a clear sense in which the rules are essential to an understanding of their activities. Whatever their underlying motives and means, they are still *playing tennis*, or *playing poker*, and as such still accept that (surreptitious infractions aside) certain rules govern their behaviour.

Similarly, it seems to me, with science. Individual scientists will have all kinds of extraneous ambitions (such as money, or fame, or prestige) and will use whatever resources they command (such as influential friends, or an established reputation, or somebody else's insights) to achieve them. But, still, the activity they are engaged in is *science*, in the sence of uncovering truths about nature, and in consequence scientists will accept that the very nature of their enterprise means that they are supposed to conduct themselves in ways that will reliably discover truths.

It might seem that the difficulty raised in this section emerges again a level up: isn't it just a matter of luck, susceptible of nothing but sociological explanation, that modern Western society should contain an institution devoted to developing reliable ways of uncovering truths? I agree this calls for sociological explanation: many societies do lack such institutions, and muddle along instead with uncritical acceptance of received doctrines, and it is by no means clear what accounts for the emergence and persistence of critical science in modern Western society. But I do not think that *this* discredits the findings of science in the slightest. On the contrary, I think we ought to celebrate our good fortune in having institutions for finding out truths.

One final caveat. It is no part of my purpose to engender dumb respect for modern scientific institutions, to assure you that the men in white coats are invariably finely-tuned truth-detectors, whose every dictum should be swallowed whole. Of course there are times where the ideology of a whole community, or the power of some group within it, gets some proposition accepted as fact despite its lack of appropriate pedigree. Of course the pressures for theories to be reliably generated do not always hold complete sway, and sometimes untruths will get written into the archives. But the appropriate response to this danger is not to belittle science as a social construction, but simply to urge and constrain scientists to be more reliable in their findings.

University of Cambridge

REFERENCES

Barnes, B. and Shapin, S. (1979) 'Darwin and Social Darwinism: Purity and History', in Barnes, B. and Shapin, S. (eds.), *Natural Order: Historical Studies of Scientific Culture*, London, Sage.

Bloor, D. (1976) *Knowledge and Social Imagery*, London, Routledge and Kegan Paul.

Bloor, D. (1978) 'Polyhedra and the Abominations of Leviticus', *British Journal for the History of Science* **11**, 245–72.

Collins, H. (1985) *Changing Order*, London, Sage.

Farley, J. and Geison, G. (1974) 'Science, Politics, and Spontaneous Generation in Nineteenth-Century France: The Pasteur-Pouchet Debate', *Bulletin of the History of Medicine* **48**, 161–98.

Goldman, A. (1976) 'Discrimination and Perceptual Knowledge', *Journal of Philosophy* **73**, 771–91.

Goldman, A. (1978) 'Epistemics: The Regulative Theory of Cognition', *Journal of Philosophy* **75**, 509–23.

Goldman, A. (1979) 'What is Justified Belief', in Pappas, G. (ed.), *Justification and Knowledge*, Dordrecht, D. Reidel.

Goldman, A. (1980) 'The Internalist Conception of Justification', in French, P., Uehling, T. and Wettstein, H. (eds.), *Studies in Epistemology: Midwest Studies in Philosophy, Volume* **5**, Minneapolis, University of Minnesota Press, 27–51.

Goldman, A. (1985) 'The Relation Between Epistemology and Psychology', *Synthese* **64**, 29–68.

Lakatos, I. (1978) 'History of Science and its Rational Reconstructions', in Worrall, J. and Currie, G. (eds.), *The Methodology of Scientific Research Programmes: Philosophical Papers, Volume* **1**, Cambridge, Cambridge University Press.

Latour, B. and Woolgar, S. (1979) *Laboratory Life*, London, Sage.

Laudan, L. (1977) *Progress and Its Problems*, Berkeley, University of California Press.

Lear, J. (1982) 'Leaving the World Alone', *Journal of Philosophy* **79**, 382–403.

Lynch, M. *Art and Artifact in Laboratory Science*, London, Routledge and Kegan Paul.

Papineau, D. (1987) *Reality and Representation*, Oxford, Basil Blackwell.

Putnam, H. (1981) *Reason, Truth and History*, Cambridge, Cambridge University Press.

JOHN F. FOX

IT'S ALL IN THE DAY'S WORK: A STUDY OF THE ETHNOMETHODOLOGY OF SCIENCE

1. SYNOPSIS

My conclusions about the work of the ethnomethodologists of science — whom, for brevity and in accordance with the verbal though not so far the published practice of other sociologists, I shall call the ethnos — are like their conclusions about the work of scientists they have studied. As they typically conclude that natural scientists misunderstand the nature of their own enterprise, or at least in their publications so write as (perhaps unwittingly) to conceal or even mislead about it, so I conclude that ethnos do. They conclude that it is mistaken to think that scientists discover about independently existing objects truths that are not artefacts of their own social and political performances. I conclude that it is mistaken to think that the ethnos have discovered or revealed such truths about scientists. I conclude in particular that they have not discovered that this traditional view of scientists is mistaken.

As an essential part of my argument I clarify some notions that are important for the ethnos' arguments and positions; in particular realism, reality, and facts. The main works I consider are an article by Garfinkel, Lynch and Livingston (hereafter GLL) on the optical discovery of pulsars; and the field's most acclaimed work and quasi-paradigm, the study of the Salk laboratories by Latour and Woolgar (hereafter LW).

2. ETHNOMETHODOLOGY

A common reaction to work that is generally called ethnomethodological — often combined with admiration — is puzzlement about the nature and motivation of the enterprise. No unequivocal resolution of this puzzle will come without careful scrutiny of the relevant texts; but these can often seem too opaque for such scrutiny. I therefore offer some preliminary (if conjectural) general clarification, without relying on it in particular interpretations.

A central idea of ethnomethodology, though rarely stated, is that it is illuminating to study the familiar in the fashion of certain old-style

Robert Nola (ed.), Relativism and Realism in Science, 59—80.

anthropologists. These used to assume that such a vast gulf was fixed between the thoughts and practices of the tribes or peoples they studied and those of their own community that the kinds of explanations that had to be offered of the former would be utterly different in kind from those appropriate for the latter. For instance, a rain-dance might be explained by the dancers as an attempt to bring rain. This can seem like much behaviour in our society in being a (less or more misguided) attempt causally to influence events. But old-style anthropologists would not seriously entertain this idea; the dance could not really be for that purpose. It should be explained rather by (what later theorists would categorize as) a *latent function* opaque to the dancers; say, as increasing social cohesion. Again, in using the same descriptive word for twins and certain birds, they would not entertain the possibility that a tribe might have a theory, with considerable explanatory potential, which indicated that these superficially different things were basically the same; rather, a prelogical (subwestern) mode of thinking must be operating.

I have cited as my examples of the traditional approach cases where it has come under effective fire, especially from Horton [cf. his (1967) and (1982)] and where it now looks absurdly patronising. But ethnomethodology has two advantages over such traditional uses of the approach. Being applied to phenomena in 'our own' culture, it is less likely to be patronising out of an assumption of vast cultural superiority (though this safeguard of course weakens as ethnomethodologists become more numerous and their subculture institutionalized in universities). More positively, since it is applied to familiar matters, the suspension of familiar assumptions and the search for quite different explanations for thought and behaviour from those which come naturally to the thinkers and behavers can turn up surprising insights.

3. GALILEANITY, REALISM AND REALITY

Notions that are central to GLL's argument are those of a *galilean* object, and of a galilean science. Verbal variants of these are central to a great deal of ethnomethodological work; indeed, the common thread of the major conclusions of ethnos' research is that the objects of science are not galilean, and that properly understood science is not. GLL provide an extended account of the notion that is a useful starting-point:

In their published article [Cocke, Disney and Taylor (1969)] the work of the optically discovered pulsar's local historicized production is rendered as the properties of an independent Galilean pulsar:

(1) The pulsar is depicted as the cause of everything that is seen and said about it.

(2) It is depicted as existing prior to and independently of any method for detecting it and every way of talking about it.

(3) The pulsar's technically detailed phenomena are made anonymous to Cocke and Disney's presence to them as witnessing persons and authors. Their presence to the phenomenon is that of Observers' Practices. We use the capitalized Observers' Practices to mean, without irony, these are practices whose technical contents are written and are readably those of a manual of disembodied proper procedures.

(4) Observers' practices in and as of the worldly *thing* are "naturalized". By that is meant (among other things) that The Observer is present to the claimed properties of the pulsar as a lucid plan of action. The Observer's actions, by reason of their compliance with the lucid plan for them, come upon the pulsar that is otherwise hidden by and unavailable to ignorance, error, vagueness, mistakes, sloppy work, careless work, circumstantial flux, etc.

(5) In the article the pulsar's identifying details are readably in the voice of the transcendental analyst. . . . (etc.) (GLL, p. 138)

This explication is crucial, because the two major things the ethnos are saying are first, that the astronomers think of and present the pulsar as galilean, and second, that it is not. The ethnos say the first clearly; that they are saying the second emerges from study of their text.

In the ethnos' definiendum ('independent galilean pulsar', or 'IGP'), 'independent' is pleonastic, and clearly the characterization is meant to be applicable to objects other than pulsars. I therefore simplify and generalize it to 'galilean object', and also simplify the definition. Their first two clauses are crucial; an object is only galilean if it exists independently of any methods of observing it or talking about it. 'Prior to' is infelicitous; the intention is surely not to exclude the Eiffel Tower from galileanity because some of the techniques by which we observe and talk about it predate its building; independence is what is crucial. Nor is it intended that a galilean object be *the* sole cause of anything, nor *a* cause of everything, that is said about it. The activities of McCallister at the telescope, as well as the pulsar, are considered by Cocke, Disney and Taylor to be an important contributing cause of the recorded observations; it is on that basis they thanked him. And some of the things that are said about pulsars — initial speculation about their theoretical possibility, expressions of doubt and so on — are not claimed by the most ardent realists to be caused at all by the pulsars. Indeed, even if an object is not a significant cause of certain *reports* of it, it may still be galilean; as the ethnos indeed imply, 'error, vagueness, mistakes, sloppy work, careless work, circumstantial flux' are categories available in the galilean framework for the downgrading of reports. That an object can be mistakenly reported does not hurt its galilean status.

Still, galilean objects are thought of as accessible in this way; the *possibility* of such causation is then essential to galileanity.

The authors also speak of galilean *science*. A galilean science is a science addressed to the study of (what it takes to be) galilean objects. But not all the objects studied and reported by a galilean science are thought of as galilean. Measurements (and experiments and observations in general) are not. Things measured usually are, properties measured frequently are. Why only frequently? Well, according to the relevant galilean theory, it will depend on the property. Suppose one is trying to measure the melting-point of silver. That the thermometer reads 980 degrees when one inspects it is not a galilean property of the thermometer; it is causally dependent e.g. on one's having stoked the furnace. That the *melting-point* is 980 is, according to the theory, not so dependent. The one reading is taken as revealing both things.

That the melting-point is thus galilean is of course a conjecture. It implies that other researchers, of varying social and political and genetic backgrounds, using a definite variety of techniques, would (barring accidents, incompetence etc.) get the same results; it implies that should silver melt naturally and someone be around to detect the temperature at which it occurred, they would (with similar provisos) get the same result.

The ethnos' definition of a galilean science applies just as well to aristotelean science, but it is worth remembering that the kind of science that derives from Galileo superseded an earlier kind that derived more from Aristotle, and which was much *more* realistic than Galileo's. What was given to experience was taken by aristoteleans to be real in what I call below the strong sense, unless tangibly abnormal circumstances obtained. Not so for the galileans; 'secondary qualities' were so given, but were not considered independent of our manner of knowing them. The truly 'galilean' objects or states of affairs were by no means manifest to the senses; for instance, the orbiting of the sun by the earth, or the rotation of the earth on its axis. Identification of the galilean objects, properties and truths through the flux of natural and contrived experience was typically a highly skilled and precarious enterprise.

4. REALITY AND REALISM

It is time to clarify further the correlative notions of reality and of realism. Realism I take as always relative to some domain of discourse. One may be realistic about tables but not about quarks, about quarks but not about sets, about energy but not about force. There are several terms for varieties of unrealism, and they must be understood as likewise relativized. One may be a nominalist about universals but not about numbers, or about properties but not about sets. One can be conventionalist about scales of temperature but not about the one-way speed of light, or about distant simultaneity but not about distant color-identity. One can be instrumentalist about quantum mechanics but not about the kinetic theory of gases, or about set theory but not about arithmetic. Realism about a domain is just the doctrine that the objects of that domain are real, or have reality; the different senses of realism are parasitic on different senses of reality. It will be enough, I think, to distinguish four.

In the first sense, objects of some kind are real just if there are entities of that kind. A second sense is rather that the truth or falsehood of claims in a domain is independent of human mental activity and of the practices involved in coming to ascertain them. It is more customary to speak of realism in the first sense with respect to kinds of objects, in the second with respect to bodies of knowledge or theory.

The second sense is independent of the first. I am an unrealist about numbers in the first sense, a realist about arithmetic in the second; I am a realist in the first sense about some experimental artefacts — there are some such things (some: for sometimes the phrase is used, legitimately, to indicate that there are not, but merely appeared to be, things of some alleged kind); that I am an unrealist about them in the second is what my calling them artefacts indicates.

A third sense of realism comes from combining the first with the second; to be a realist about a domain in this, the *strong* sense, is simply to be a realist about it in both of the first two senses.

A fourth sense of realism is a weakening of the third. Something is real in this sense if it is real in the first, and confronts individuals (rightly) as existing independently of their thoughts or perceptions or volitions. The key word is 'individuals'; dependence on thoughts, perceptions and volitions in general is not being excluded. Sometimes 'social' or 'cultural' is used, by contrast with 'natural' or 'physical', to express this sense of reality.

Multiplication of senses is only necessary where ambiguities have been causing trouble. So I will not speak of a distinct sense of 'realism' for the claim that the objects concerned play a causal role, in particular a causal role in our knowledge of them. Still, at times this claim is crucial, so when it is made, I shall call the whole position 'causal realism', realism being understood in one of the senses already explicated.

So to consider an object galilean is to be a causal realist about it in the third sense; to believe in a galilean science is to be such a realist about its purported objects.

Sometimes unrealists (in one or another sense) find language which in its ordinary sense expresses realism so convenient that they adopt it, explaining elsewhere (sometimes none too clearly) that they do not intend the realistic connotations. Thus I am prepared, doing arithmetic, to answer the question 'How many primes are there between ten and twenty?' with 'Four', and doing metaphysics, to explain that I do not believe in numbers at all. The onus is then on me to offer a reductive account of what I really meant by what looked like an existence claim. So too Marx, doing technical economics, insisted on the exchange-value of goods, and in his wider social analysis denied that goods had any such property; claimed that what 'appeared' to be such a property was to be analysed in terms of social relations among people and classes; and even 'classes' were in turn described as 'illusory', though they were assigned causal efficacy! Perhaps some clarity in the interpretation of Marx can be achieved if I offer also this definition (by analogy with that of 'galilean'): to consider an object *marxian* is to be a causal realist about it in the fourth sense but not in the second.

5. THE UNREALISM OF GARFINKEL, LYNCH AND LIVINGSTON

GLL repeatedly reject strong realism about the pulsar; what the astronomers display as galilean characteristics are in fact purely the result of their own work:

The IGP . . . is a *cultural* object, *not* a "physical" or "natural" object.

> Our insistence that the IGP is a cultural object insists upon this about it: astronomically detailed specifications exhibit as the pulsar's galilean independence the local production properties of Cocke and Disney's work. It is the locally produced and locally recognized orderliness of Cocke and Disney's embodied practices as of which the exhibitable objectivity and the observable analyzability and intelligibility of the phenomenon's technical, identifying details consists — definitely, exactly, only, and entirely. (GLL, p. 141)

Why do they reject the galilean account? The first reason they offer is that the report in which this account is presented is misleading:

> Their mathematical collection of equivalent observations is . . . misleading . . . because it is . . . prominently obvious in their tape and log, but noticeably absent from their article, that their collection, when it is examined in the light of first time through, was obtained, and was only obtainable, case-after-case, as an historicized series. The series was done as a lived orderliness, in real time. Only as a feature of its local historicity did the series project as its possibility that it could become an atemporalized collection of measurable properties of pulse frequency and star location that according to a Galilean science are independent of the local practices as of which just this gathering of observations was composed. (GLL, p. 135)

In what respects is the astronomers' report here claimed to be misleading?

Could it mislead its readers into thinking that the collection of observations was *not* obtained as an historicized series, was *not* done as a lived orderliness, or not in real time? I am not sure exactly what would count as an unhistoricized series of observations, done as an unlived orderliness or in unreal time. Still, I conjecture that *no* reader of the astronomers' article was in peril of being so misled. Even had the authors, alerted by ethnos to such peril, included a prophylactic disclaimer, confessing the historicized nature of the series of their observations, the livedness of the orderliness as which and the reality of the time in which they were done, surely no reader would have been a jot the wiser.

The crux of the ethnos' argument is that only as a feature of its local historicity could the series be construed in terms of a galilean object. Only if people at definite times and places, on definite occasions, record observations of a star, can a chart be produced of its positions. Or as the ethnos say, 'Only as a feature of its local historicity does a series project as its possibility that it could become an atemporalized collection of measurable properties of star location that according to a Galilean science are independent of the local practices as of which just this gathering of observations was composed.' But the most galilean of readers would not have doubted it. The ethnos are mistaken in finding the report misleading for the reasons they adduce.

They also present a second, strategic or methodological reason for insisting that the pulsar is a cultural object, only and entirely locally produced.

> We now explain that policy.
>
> The "potter's object" is at the heart of that policy. We're going for a reversal of the conventional distinction between the real pulse and the apparent pulse. According to the conventional distinction the real pulse is the presence of the pulse to be derived from the characteristics of the apparent pulse when those apparent characteristics are subjected to careful inference and test. Our notion of the pulsar locates it at hand as *that which is real in and as of inquiry's hands-on occasions.* We want the real pulsar to become available to Cocke and Disney in and as of their primordial embodied practices of finding-and-exhibiting-again an IT's adequately astronomical technical details. (GLL, p. 137)

GLL are commendably open about their strategy; it is worth becoming clear about its implications. They are resolved to call 'the real pulsar' only what is immediately accessible to the scientists, only what is locatable *in* their practices. So they are resolved not to apply the phrase to anything inferred by argument and test from what is so accessible. Whatever they identify as 'the pulsar' or 'the object', that it has this feature is decided *a priori*. That it is not galilean is not a conclusion of their research but a defining feature of their strategy. It is simply an artefact of the strategy, and depends in no way on one thing rather than another happening in the laboratory. They insist not only on the equation of the 'optically discovered pulsar' with 'the night's work' of the scientists (rather than with anything transcending that work with which the work put them in touch), but that this equation is a matter of their stipulation of meanings:

> "(ODP)", "the night's work of the optically discovered pulsar" and "the night's work" are used synonymously and interchangeably. (GLL, p. 141)

Nor is it even claimed that their interpretation explains matters better than the alternative. On the contrary. The alternative is deliberately to be disregarded and not taken seriously:

> ... Our policy, and the point: We want to examine the pulsar for the way it is *in hand* at all times in the inquiry. We want to see the way it is "performatively" objective.

We did *not* examine and we want *not* to examine the end-point object for its correspondence to an original plan. We want to disregard, we want *not* to take seriously, how closely or how badly the object corresponds to some original design — particularly to some cognitive expectancy or to some theoretical model — that is independent of their embodied work's *particular occasions* as of which the object's production — the *object* — consists, only and entirely. (GLL, p. 137)

But what kind of object do GLL claim the pulsar is? The ethnos' story is that the pulsar is not galilean, but locally produced; not a physical, but a cultural object. But what on earth is this cultural object?

Different understanding [sic] of the IGP as Cocke and Disney's discovery need to be sorted out. A first and incorrect understanding takes the Independent Galilean Pulsar to be an *account*. The IGP can be construed as an achieved account of their night's work. . . . But taking the IGP as an account can be irreparably distracting. It needs to be remembered that Cocke and Disney don't discover an *account* of the pulsar; they discover an astronomically demonstrable pulsar. Correctly understood, the IGP is an *object*, and a cultural object, not a "physical" or "natural" object. In the entirety of its technical astronomical properties, IGP is a cultural object.

That object, the IGP, is *not* an account.

That object makes their night's work account-able. That object, the IGP, makes witnessable and discourse-able the apt efficacy of their practices for locating, finding, collecting, demonstrating its adequately-astronomically-technical-details-again. Their actions are rendered, in the properties of the IGP, as pulsar-demonstrating astronomically analyzable and intelligible practices. (GLL, p. 141—2)

Several puzzles arise from this text, of which I shall consider only a few. Firstly, why on earth would this be a *first* understanding? Neither layperson nor astronomer would spontaneously be tempted to mistake a pulsar for an account of one, nor vice versa. To speak of a *galilean* pulsar, an 'IGP', should make the avoidance of this mistake doubly sure; for the point of the qualification is to ensure that the word 'pulsar', so qualified, connotes a physical object whose existence and attributes are independent of the work of its observers. The only people I know who seem to have found this a temptation are LW, who claim that the chemical TRF(H) ["thyrotropin releasing factor (hormone)"] and *statements* ostensibly about it are 'the same thing'. [LW, p. 177] Perhaps the answer is this: if one insists that the scientists' object was only and entirely something locally produced by them, it is hard to see what it could be if not their 'account'; if one insists also on calling it a pulsar, one could end up describing a pulsar as an account.

Secondly, why did the authors choose to express their thesis in a contradictory form? For in insisting that the *IGP* is not a physical but a cultural object, the authors literally contradict themselves; their characterization of galileanity was devoted precisely to elucidating what it was to be a physical rather than a cultural object. Perhaps they mean simply to say that the pulsar Cocke and Disney discovered, which as astronomers they *took* to be a physical, galilean object, is really a cultural object. For this is what they imply elsewhere: what the astronomers 'exhibit as' the pulsar's galilean independence, the ethnos 'insist', is really something 'locally produced'. Since otherwise their position is trivially inconsistent, I shall assume that they do mean this. Still, why the inconsistent phrasing?

Perhaps their reasons have to do with respect and tolerance for the subjects of their study. Ethnos generally do not want to buy into controversies with the subjects of their studies on their own ground; indeed, there are constant gratuitous affirmations of their subjects' competence, even of the demonstrated character of their conclusions. What to do, then, when the ethnos' interpretation of something (the nature of what Cocke and Disney identified, say) clashes with the scientists'?

For they *do* clash. The astronomers claim their observations are of a (galilean) pulsar, but the very features they 'exhibit' as 'the pulsar's galilean independence', according to the ethnos, are rather those of a cultural object that is 'only and entirely' 'locally produced'.

Well, one strategy for neutralizing this conflict is to *redefine astronomy*. What the astronomers *say* is deemed legitimate, but it is *reinterpreted* by the ethnos so that it no longer clashes with what the ethnos wish to say; the astronomer's interpretation of what they say, insofar as it still so clashes, is treated as an untutored incursion into *sociology* and so as a pardonable blunder. (In this case, the astronomers' interpretation of their observations as caused by a pulsar understood as a physical object with characteristics independent of their own observing practices is the blunder, to correcting which the entire article is devoted.)

But the ethnos have an implicit principle of tolerance. It recalls those of Reichenbach and Carnap: they allow their subjects to *say* what they want, as long as they don't mean what the ethnos wouldn't mean. In particular, their claim that they discovered a pulsar is *granted*; but since the ethnos want to disregard whatever is not locally produced, they postulate such a 'cultural object' and call it the pulsar the astronomers discovered. The ethnos had introduced the phrase 'independent galilean pulsar' precisely for the astronomers' rival conception of their discovery. But noticing that what they meant by 'IGP' was precisely what the scientists meant by 'pulsar', they confusedly allowed them even that phrase, though it had not been the scientists', but their own term of art intended to differentiate the scientists' interpretation from their own.

But none of this answers the question, of what kind do the ethnos suppose is this 'pulsar' that is a 'cultural object' but not an account?

In the course of their night's work, the astronomers referred frequently to 'it', as if they were referring to the one thing all along. They described 'it' as in the middle of the screen, as growing up the side a bit, as having moved to the right, as being a pulse, as being possibly an artefact, as being better than it was last time, and finally, as being a bloody pulsar. Now what would be in the middle of screen, or growing up the side of the screen, is a visual image. Images, too, not (galilean) pulsars, increase in size when telescopic magnification is increased, shift to the right of the screen when the telescope is shifted to the left. Some of what they say about 'it' can apply to the image, some — particularly after the pulsar has confidently been identified — to the pulsar; but not all they say can consistently apply to both.

The realist interpretation of this discourse is that 'it' does not maintain constant reference; that 'it' refers sometimes to the image, sometimes to the (galilean) pulsar.

Whether this interpretation is right or wrong is a question of astronomy, but its *imputation* is surely correct *sociology*; it is what the scientists intend. (Further semantic and ontological complications can here be ignored: for instance, in saying that an image has shifted, one is arguably equating two successive images, on the assumption of significantly common causality.)

This is how GLL summarize such facts:

> . . . by reason of first time through, the work of the optically discovered pulsar . . . "evolved: from an evidently-vague IT which was an object-of-sorts with neither demonstrable sense nor reference, to a "relatively finished object". (GLL, p. 135)

The description of the discourse as unitary in reference, and so as evolution of an 'evidently-vague IT' to a 'relatively finished object', is question-begging against realism; for if there is a (galilean) pulsar out there in space, *it* did not evolve during the night from 'an evidently-vague IT'.

But what, again, is their alternative account of the pulsar? It might be considered unkind to conjecture of them what Cardinal Manning claimed of Herbert Spencer — that they have no mental equivalents for their terms, none whatsoever — but in fact they proclaim this themselves, as clearly as they proclaim anything:

"Sense and reference" as an analytic apparatus is unavailable, distracting and even useless until the relatively finished object is at hand. And then the apparatus is available in that it is grounded in the *Galilean object* — not in the (ODP). (GLL, p. 158)

So GLL claim there is no coherent account of what 'it' refers to till the scientists have reached the stage of talking simply about a galilean object, and then it is the *galilean* framework that supplies it. (I would rather say that there is no coherent *unitary* account of 'it''s reference throughout the night's work, but that before a galilean object is identified the *image* provides a reference.) GLL are right in admitting that no coherent sense can be made or reference supplied for the scientists' discourse in terms of unitary reference to a nongalilean object. So much the worse, though, for their insistence that this is how the discourse must be construed; that 'IT' is *all* in 'the night's work'.

Such 'evolution', by the way, is not, as GLL suggest, peculiar to situations of 'first time through'; it happens routinely. Tuning a television set to a cricket match, one might say successively 'It's just a blur', 'it's still wobbling horizontally far too much in the middle of the set' and 'Oh, it's Martin Crowe in the slips'. Someone very ignorant of television or of English usage might be confused into thinking that Crowe was just a blur (and so not a well-defined person), or guilty of wobbling too much horizontally; though it would take ethnomethodologists to decide that Crowe fielding in the slips existed only and entirely in the 'hands-on practices' of the knob-twiddler.

6. THE UNREALISM OF LATOUR AND WOOLGAR

LW acknowledge that the central connotation and point of the term 'fact', for their subjects in their native thickets, is galilean:

We presented the laboratory as a system of literary inscription an outcome of which is the occasional conviction of others that something is a fact. Such conviction entails the perception that a fact is something which is simply recorded in an article and that it has neither been socially constructed nor possesses its own history of construction.

Like GLL, they co-opt the term for something about which they proclaim an anti-galilean thesis. They continue:

Understanding the nature of a fact in these terms would obviously hinder any attempt to implement what has been called the "strong programme" in the sociology of science. In this chapter, we shall attempt to examine in detail how a fact takes on a quality which appears to place it beyond the scope of some kinds of sociological and historical explanation. In short, what processes operate to remove the social and historical circumstances on which the construction of a fact depends? (LW, p. 105)

What is their new understanding of the term 'fact'? Well, to what sort of thing do they apply it? In the first instance, it seems, to whatever scientists apply it to (though the ethnos reconstrue such 'facts' so drastically that in the second instance a claim of coreference becomes implausible):

We have chosen to study the historical genesis of what is now a particularly solid fact. TRF(H) is now an object with a well-defined molecular structure, which at first sight would hardly seem amenable to sociological analysis. If the process of social construction can be demonstrated for a fact of such apparent solidity, we feel this would provide a telling argument for the feasibility of the strong programme in the sociology of science. In short, our objective in studying the genesis of TRF . . . (LW, p. 106—7)

It seems that the apparently solid fact whose genesis they are studying is TRF itself. But there are reasons for hesitation. Gold, like TRF, is taken by scientists to be a fact, with a definite microstructure. And LW are not arguing that among the things scientists take for facts, some are and some aren't; they accept the scientists' *extension* for 'fact' and offer a revisionist account of *what it is to be* a fact. TRF and gold are, they would say, both facts. But facts are, despite the illusion of out-thereness and of objectivity to which scientists succumb, socially constructed. Out of what? Out of *statements*:

We shall specify the precise time and place in the process of fact construction when a statement became transformed into a fact and hence freed from the circumstances of its production. (LW, p. 105)

The collection of words acquires a currency in discourse which creates a referent in the real world. The real worldly object is thus created by virtue of the statement. [Woolgar (1981a) p. 384]

The reason for this hesitation is simply the implausibility of this claim. If gold could really be constructed out of statements, and LW have worked out how it is done, they would be richer than they are. Barnes made a similar point:

With cream-cakes there is a chance of satisfying hunger — with accounts of cream-cakes there is not. . . . Truly, if this were the case, there would be no need for the capitalist mode of production: talk would indeed be work. [(1981) p. 492)

Woolgar's reply (much shortened) is instructive:

. . . to speak of "accounts of cream cakes" as different from "cream cakes" is to impute an objectively independent existence to the latter entity, however much one would wish to claim that the "account" is the product of social circumstances . . . [Such views] contrast significantly with the view that accounts are *constitutive* of reality . . . [which view] provides a powerful analytic handle. For it . . . enables us to ask how in practice [those who distinguish accounts from what they are accounts of] manage to ignore or evade the implications of the position that accounts are constitutive of "reality". . . . bakeries, the capitalist mode of production, interests and the like . . . are just . . . accounts . . . [(1981b) p. 507–8]

So he reiterates that entities are accounts, and acknowledges that to speak of cream cakes as different from accounts of cream cakes conflicts with the 'constitutive' view he has defended. Still, he insists that of course he does not deny this obvious distinction! [(1981) pp. 506, 508]

Just who is it, then, who manages "to ignore or evade the implications of the position that accounts are constitutive of 'reality'"?

Do ethnos 'really mean' something more moderate; perhaps that the stuff that is called TRF or gold is galilean, but that its status among scientists as a fact depends on laboratory work and social factors? LW repeatedly rule out such a 'moderate' reading of their text:

. . . in emphasizing the process whereby substances are *constructed*, we have tried to avoid descriptions of the bioassays which take as unproblematic relationships between signs and things signified. Despite the fact that our scientists held the belief that the inscriptions could be representations or indicators of some entity with an independent existence "out there", we have argued that such entities were constituted solely through the use of these inscriptions. It is not simply that differences between curves indicate the presence of a substance; rather the substance is identical with the perceived differences between curves. (LW, p. 128)

So they reject the scientists' basic claim that the inscriptions even could be indicators of some independent entity. Again, the scientists consider that TRF existed before the research into it was done; that what the research achieved was the skilful revelation of its existence and structure.

On the other hand, what of the epistemic or social status of these claims, as claims that are now solid by current scientific standards and can, barring new surprises, reasonably be taken for granted in further research? Scientists are under no 'illusion' that this status is independent of scientific practice. It is for practice on which such status is recognized as depending that scientific honours are awarded. So it is not this status, but the galilean reality of TRF, that LW are here attacking.

> ... despite these [convincing] arguments, facts refuse to become sociologized. They seem able to return to their state of being "out there", and thus to pass beyond the grasp of sociological analysis. In a similar way, our demonstration of the microprocessing of facts is likely to be a source of only temporary persuasion that facts are constructed. Readers, especially practising scientists, are unlikely to adopt this perspective for very long before returning to the notion that facts exist, and that it is their existence that requires skillful revelation. . . . As Kant advised, it is not enough merely to show that something is an illusion. We also need to understand why the illusion is necessary. (LW, p. 175)

About facts they are realists in the first and at most the fourth, but not in the second sense:

> We do not wish to say that facts do not exist nor that there is no such thing as reality. In this simple sense our position is not relativist. Our point is that "out there-ness" is the *consequence* of scientific work rather than its *cause*. (LW, p. 182)

What, at last, are facts according to LW, and what reasons do they offer for their own and against a galilean account?

They offer three conflicting accounts of what facts are. The first describes facts simply as things or as stuff; they call TRF a fact and accept the scientists' description of it as a white powder. The second describes facts as perceived differences between what scientists take as mere indicators of facts, and describes TRF as the perceived difference between curves. The accounts conflict; it is observably false that the perceived difference between curves is a white powder. The third account equates facts with statements. (LW, p. 177)

Their arguments against strong realism come in the context of argument for their second and their third accounts of facts. For their second:

> . . . an object can be said to exist solely in terms of the difference between two inscriptions. In other words, an object is simply a signal distinct from the background of the field and the noise of the instruments. (LW, p. 127)

This looks like an inference from the conditions under which, as observers of laboratory practice, they have noticed an existence-claim is taken as *warranted* ("can be said . . .") to what it must be (despite the interpretations of the practitioners) that is claimed to exist. If so, it is an application of the verification theory, that the meaning of a claim consists in the conditions under which it is taken as warranted. This is sufficient criticism.

Their arguments for equating facts with *statements*, on the other hand, consist simply in arguments against what by then they treat as the sole alternative, the (galilean) 'realist' account. The first argument is that realists cannot describe the 'out there-ness' of objects without simply reformulating statements that purport to 'be about' such reality, e.g. that TRF is Pyro-Glu-His-Pro-NH_2.

This is correct, but worthless as argument. It is a general truth that one cannot describe objects as being of a certain kind without producing statements that purport to be about objects of that kind, and if this were sufficient to discredit such descriptions it would be sufficient to discredit all discourse.

LW also argue from the existence of artefacts, from the process of

> *deconstruction* of reality. The reality "out there" melts back into a statement, the conditions of production of which are again made explicit. (LW, p. 179) . . . The importance of observing the transformation of a statement between fact-like and artefact-like status is obvious: if the "truth-effect" of science can be shown both to fold and unfold, it becomes much more difficult to argue that the difference between a fact and an artefact is that the former is based on reality while the latter merely arises from local circumstances and psychological conditions. The distinction between reality and local circumstances *exists only after* the statement has stabilized as a fact. (LW, p. 180)

This is an interesting argument, analogues of which seem widely influential. But it has two strange features.

First, the data it appeals to seem rather to support the realist position. The 'folding and unfolding' of the 'truth-effect' is this: scientists think of a phenomenon (the obtaining of some regularity) as perhaps depending on local conditions. Finding that it persists under considerable variation of these, they come to suspect it of being after all a (galilean) fact; finding again failures of such persistence and/or plausible local explanations, they demote it again to artefactual status.

The scientists see themselves as trying to identify galilean facts, and take persistence and failure of persistence under variation of local conditions as conjectural evidence for and against such factuality. Thus on the realist story such folding and unfolding is explained and to be expected.

Second, the antirealist account seems unable even coherently to describe the situation. If 'reality' is taken in the strong sense, and no other is relevant, the distinction between reality and the local operations of scientists cannot by definition depend on the stabilizing of a statement. For consistency, all LW can mean is that the two are only distinguished *by scientists* after such stabilization. This is not incompatible with realism. But it is mistaken. First, the distinction *is* conjectured beforehand; second, its being made confidently is just what such stabilization *is*.

7. FACTS

Looking at different usages of 'fact' may help sort out some of LW's argument. One thing implied in calling something a fact is that it is true; indeed, the simplest usage of 'a fact' is simply as a synonym for 'true'. (Is that a fact?)

A second usage of 'fact' is strongly realistic; according to this usage, a fact is not only something true, but whether or not something is a fact does not depend in the least on what we think or know about it, or on our research practices. To show this latter dependence is to show that a claim has not factual, but artefactual status.

A third usage of 'fact' contrasts facts with *things*. Facts are rather understood as *ways things are*. Thus Wittgenstein insisted that the world was the totality of facts, not of things. Understanding this sense helps clarify some obscure and confusing fact-talk.

A central feature of such talk is that the criteria used for identifying facts are parasitic on the criteria used for identifying propositions. That is, firstly, each instance of the following schema is taken as true: The fact that p is identical with the fact that q just if the proposition that p is identical with the proposition that q. And secondly, it is such ways as we have (often described as 'ascertaining meaning') for deciding the identity or distinctness of propositions that we use in order to ascertain that of facts. There are no independent ways of ascertaining that of facts that could be used to help us with the question about propositions;

this is a verbal point about the usage of 'fact'. For instance, when we learn that it is the same thing that is denoted by the phrases 'The Evening Star' and 'Venus', we thereby learn that the transit of Venus is the same state of affairs as the transit of the Evening Star. But 'fact' differs in usage from 'state of affairs' precisely in this: we may not conclude that they are the same fact. We may speak of someone unaware of the co-reference as knowing the fact that Venus was in transit, but not the fact that the Evening Star was.

The third usage can be explained as growing out of the first. Such statements as "I am aware of the fact that you dislike artichokes" can be plausibly construed simply as "I know (it is true) that you dislike artichokes". Most such fact-statements are entirely explicable in terms of the first idiom. Others can be explained on the *basis* of it. Either by linguistic confusion, or through a metaphysical postulate that what *makes* truths in the first sense true are 'truths' or 'facts' 'in the world' to which they so exactly 'correspond' that they can be named by the same phrases (e.g. "the fact that you dislike artichokes") a world of such entities is postulated. It is even, as by Wittgenstein and Austin, confused or identified with the world, period.

This *metaphysic* of facts is not to be confused with other variants of 'correspondence theories of truth', some of which postulate *things*, or even states of affairs (which do not have the identity-criteria of 'facts') to render truths true. Few believe in 'facts' in this third sense; Prior, [(1971) Ch. 1] lucidly defends such unbelief. Strawson (1950) expresses it gnomically:

> Of course statements and facts fit. They were made for each other. If you prize the statements off the world you prize the facts off it too; but the world would be none the poorer.

This idea of 'fact' and scepticism about it can make sense of LW's remarks about their identity with statements:

> . . . we offer our observations of the way this kind of illusion is constructed within the laboratory. It is small wonder that the statements appear to match external entities so exactly: they are the same thing. (LW, p. 177)

However, such sense as it makes depends on the sharp contrast between facts and things, and so is lost when they start describing white powders as statements.

There is a fourth usage of 'fact', in which the principal connotation is general acknowledgement (by some relevant community) of some truth being reasonably 'established'. Among the categories with which that of 'fact' is being contrasted here are not merely 'error' but 'speculation' and 'plausible hypothesis'. I ignore as here irrelevant several further related usages, e.g. that of 'fact' by contrast with (even 'well established') *theory*.

What LW take their laboratory studies to 'demonstrate' and 'show' is the nongalileanity of facts in the first sense, i.e. the non-existence of facts in the second. What they do show is the nongalileanity of factual status in the fourth sense. But this is trivial; of course it depends on the activity of humans whether it is a fact, i.e., a *generally acknowledged* truth, that salt is soluble in water. It does not follow that it depends on the activity of humans whether it is a fact in the first sense, i.e., whether salt is soluble in water. So their 'demonstration' of the social construction of facts fails.

8. THE ARTEFACTUALITY OF ETHNOMETHODOLOGICAL THESES

We saw that GLL's conclusions were artefacts of their 'potter's object' strategy, and that this was not even concealed, except by their prose style. I shall argue that so are LW's conclusions, and shall show that exclusive use of ethnomethodology guarantees such conclusions.

Sometimes ethnos proclaim that ethnomethodology alone is legitimate; that other approaches are obscurantist or unscientific; that to share an explanatory category or assumption with the subjects of ones study is to 'go native'. LW count as moderates:

> It is not necessary to attach any particular significance to the achievement of a "correct" balance between "social" and "intellectual" factors. This is for two main reasons. Firstly . . . the distinction between "social" and "technical" factors is a resource drawn upon routinely by working scientists. . . . Secondly . . . we regard the use of such concepts as a phenomenon to be explained. More significantly, we view it as important that our explanation of scientific activity should not depend in any significant way on the uncritical use of the very concepts and terminology which feature as part of that activity. (LW, p. 27)

Their moderation consists in their presenting their maxim as theirs rather than as a general imperative, and in their use of the qualifier 'uncritical'. But the qualification is cosmetic. In practice they consider that the only and necessary alternative to 'uncritical' acceptance or use is complete eschewing.

The radical nature of this alternative should not be underestimated. Scientists and layfolk may well offer as part of the explanation of what scientists do and say that certain substances have properties which become manifest under experimental conditions; indeed, that such conditions are typically contrived just in order to become acquainted with such properties. So to admit any part of this explanation would be to *go native*, would be ruled out *a priori* by the resolve to stick solely to ethnomethodology. Such resolvers I shall call *ethnocrats*.

Ethnocracy has provided some of the motivation of what has been called 'discourse analysis'. Aware that scientists hold all sorts of beliefs about the world they study, that they use some of these in explaining what happens in their laboratories, and that the scientifically literate can hardly help 'falling into' sharing some such beliefs, some scrupulous ethnocrats have concluded that sociologists and historians should concern themselves exclusively with the *utterances* produced by scientists and the relations among them; only avoiding overlap of subject-matter is a strong enough prophylactic against the crime of overlap of explanatory beliefs.

This is not the only motivation of 'discourse analysis'. A strong scepticism about the study of science, combined with ultrapositivist ideals, are others. Some theorists think it is *impossible* to give an account of what actually occurs in science, apparently because of conflicts between competing accounts, yet that it is possible to answer definitively questions about the analysis of discourse; 'the analyst is no longer required to go beyond the data'. (Mulkay & Gilbert, p. 314)

As with ethnomethodology in general, discourse analysis can of course be useful as a method; I am commenting here on its ethnocratic advocacy.

Ethnocracy and ultrapositivism only partly explain the popularity of extreme antirealism. There is also the 'more radical than thou' phenomenon. Among the left in the '60s there was no discernible upper bound to the idiocy of stances that might be adopted under pressure to show oneself more radical than some rival faction. Some of the pressure has moved into sociology with some of the personnel. But this explanation is more relevant to camp-followers than to the recognized leaders I have selected for analysis here.

9. THE REFLEXIVITY PROBLEM

As Bacon saw, error can be fertile; even foolish error. But conversely, not everything with a chance of fertility fails to be foolish error; and ethnocracy is a case in point. If the explanation of behaviour offered by the behavers is ever correct, an ethnocrat is precluded from getting that explanation right. Nor can the goal of producing a general scientific account of science ethnomethodologically be in principle achieved. For any account that applies to itself shares explanatory perspectives and categories with at least one subject of its study, and so fails in ethnomethodological character. And any account that fails to apply to itself fails either in generality or else in scientific character.

Because of this proof, some determined ethnocrats reject the "strong programme"'s demand that the sociology of science be reflexive, i.e. the demand that an adequate explanation of scientific behaviour should itself be scientific, and so apply to itself. Collins [(1981) p. 216, (1982) pp. 140—1] argues that scientists in general study things and produce talk about things, talk distinct from the things they study; so to mimic scientists accurately sociologists should do likewise; so the demand for reflexivity is a demand to deviate from scientific practice, and should be rejected.

Such ingenuity is worthy of a less perverse cause. As well might logicians claim licence to perpetrate fallacies in works *on* logic.

Others merely forgo generality as an ideal, offering only such vague metageneralities as that no two cases of supposed scientific success are at all comparable; or, like some discourse analysts [cf. Mulkay (1981) 170], forgo explanation as an ideal. But is not this in effect to abandon the sociology of science, however fascinating the preferred alternative? At least this has been shown: reflexivity is required for any general, scientific, explanatory sociology of science, and ethnomethodology cannot be reflexive.

La Trobe University

REFERENCES

Barnes, B. (1981) 'On the 'Hows' and 'Whys' of Cultural Change' (Response to Woolgar), *Social Studies of Science* **11**, 481—98.

Cocke, W. J., Disney, M. J. and Taylor, D. J. (1969) 'Discovery of Optical Signals from Pulsar NP 0532', *Nature* **221** (Feb. 8).

Collins, H. M. (1981) 'What is TRASP? The Radical Programme as a Methodological Imperative', *Philosophy of the Social Sciences* **11**, 215—224.

Collins, H. M. (1982) 'Special Relativism — The Natural Attitude', *Social Studies of Science* **12**, 139—143.

Garfinkel, H., Lynch, M. and Livingston, E. (1981) 'The Work of a Discovering Science Construed with Materials from the Optically Discovered Pulsar', *Philosophy of Social Science* **11**, 131—158.

Horton, R. (1967) 'African Traditional Thought and Western Science', *Africa* **37**; rp. (abridged) in Wilson, B. (ed.), *Rationality*, Oxford. Blackwell.

Horton, R. (1982) 'Tradition and Modernity Revisited', in Hollis, M. & Lukes, S. (eds.), *Rationality and Relativism*, Oxford: Blackwell; pp. 201—260.

Latour, B. and Woolgar S. (1979) *Laboratory Life: The Social Construction of Scientific Facts*, Beverly Hills, Calif.: Sage.

Mulkay, M. J. (1981) 'Action and Belief or Scientific Discourse? A Possible Way of Ending Intellectual Vassalage in Social Studies of Science', *Philosophy of Social Science* **11**, 163—171.

Mulkay, M. J. and Gilbert, G. N. (1982) 'What is the Ultimate Question? Some Remarks in Defence of the Analysis of Scientific Discourse', *Social Studies of Science* **12**, 309—319.

Prior, A. N. (1971) *Objects of Thought*, Oxford: Clarendon.

Strawson, P. F. (1950) 'Truth', *Proceedings of the Aristotelean Society*, Supplementary Vol. **XXIV**.

Woolgar, S. (1981a) 'Interests and Explanation in the Social Study of Science', *Social Studies of Science* **11**, 365—394.

Woolgar, S. (1981b) 'Critique and Criticism: Two Readings of Ethnomethodology', *Social Studies of Science* **11**, 504—514.

PHILIP PETTIT

THE STRONG SOCIOLOGY OF KNOWLEDGE WITHOUT RELATIVISM

0. INTRODUCTION

Under the more or less established picture of the discipline, the sociology of knowledge — if you prefer, of received opinion — eschews any form of relativism. It distinguishes between knowledge proper and mere ideology and it seeks only to give a social explanation of the claims made by the latter. At the cost of having to deploy such a controversial distinction, it avoids any suggestion that serious cognitive claims — in particular, those of respectable science — are a function of local context. On the contrary, it suggests that certain claims may enjoy absolute merits, transcendent of their context of origin. [For a contemporary version see Laudan (1977).]

In recent years the sociology of knowledge has broken with this tradition of respect. It has been presented, and indeed pursued, under the guidance of the so-called strong programme. This involves four constraints on the sociology of knowledge, which have been more or less canonically formulated as follows (the passage is from Bloor (1976), pages 4—5):

1. It would be causal, that is, concerned with the conditions which bring about belief or states of knowledge. Naturally there will be other types of causes apart from social ones which will cooperate in bringing about belief. [Also see Bloor (1981), page 199.]
2. It would be impartial with respect to truth and falsity, rationality and irrationality, success and failure. Both sides of these dichotomies will require explanation.
3. It would be symmetrical in its style of explanation. The same types of cause would explain, say, true and false beliefs.
4. It would be reflexive. In principle its patterns of explanation would have to be applicable to sociology itself. [Also see Barnes (1974).]

The strong programme is clearly very attractive. It is independent of the sort of controversial distinction which the weak programme, as we might call it, is forced to deploy. It has the hallmark of an empirical,

Robert Nola (ed.), Relativism and Realism in Science, 81—91.

open-minded plan of work. And it holds out the promise of iconoclastic results as the claims of scientists, so often ascribed almost vestal purity, come within the embrace of the sociologist. [See Bloor (1976), Chapter 3.]

The only feature of the strong programme that might give one pause is that it is taken by its defenders to entail a full-blooded relativism [See Barnes (1974), page 154; Bloor (1976), page 142; Barnes and Bloor (1982).] In this respect, as in those which have a more immediate appeal, it is held to be the contrary of the weak programme that it sought to replace.

The question with which I am concerned in this paper is whether we can espouse the strong programme for the sociology of knowledge without necessarily committing ourselves to relativism. Can we have the best, as many will see it, of both worlds? I shall try to establish, against the defenders of the programme, that we can.

My paper is in three sections. First I look at the relativistic image of knowledge, in particular scientific knowledge, that is associated with the strong programme. Next I show that the tenets of the programme can and should be interpreted so that such relativism is not entailed. And finally I argue that it is an independent commitment to a sort of conservatism which best explains the relativistic inclinations of defenders of the programme.

1. THE RELATIVISM OF THE STRONG PROGRAMME

Barnes and Bloor (1982) argue that relativism about beliefs on any topic is motivated by the observation that those beliefs vary widely — and vary so as to generate conflict — between individuals, schools, periods, cultures, or whatever. But such variety does not amount to relativity, as they admit, since it may still be the case that at most one of the possible sets of beliefs is true. Relativism begins to appear at a second stage when, in seeking to explain the variety, one asserts that what people believe is determined by their local context. Yet even such context-related variety falls short of relativity, as they again acknowledge, since it may be that one of the contexts is more suitable than others for the generation of true beliefs. Relativism only comes properly on the scene when one denies that any contexts are superior in this regard.

This last, crucial claim comes to what Barnes and Bloor describe as an equivalence thesis: a thesis to the effect that there is no difference of quality between the situations, and certainly between the beliefs, of contending parties. One might seek to express the thesis in the claim that all beliefs are equally true or equally false but many paradoxes lie along that way. Barnes and Bloor claim to find a more satisfactory expression of this equivalence in the postulates of impartiality and symmetry.

All beliefs are on a par with one another with respect to the causes of their credibility. It is not that all beliefs are equally true or false, but that regardless of truth and falsity the fact of their credibility is equally problematic. [Barnes and Bloor (1982), page 23.]

It is admitted that those who believe certain things will usually have reasons to offer for their beliefs and will always regard their beliefs as true. It is admitted furthermore that someone — for our purposes, the sociologist — who seeks to explain a set of beliefs will usually hold a position on their rationality and truth-value. What is maintained in this expression of relativistic equivalence is that the assessment has no bearing on how the sociologist ought to go about explaining the beliefs, and that equally the explanation he endorses has no relevance to how he ought to assess them.

The fact that the sociologist thinks the beliefs are true or rational, for example, does not give him any reason to think that questions of explanation should be answered one way rather than another.

All of these questions can, and should, be answered without regard to the status of the belief as it is judged and evaluated by the sociologist's own standards. [Barnes and Bloor (1982), page 23.]

And the fact that he thinks certain beliefs are explained by factors usually associated with falsehood and irrationality, for example, does not give the sociologist a ground for making an unfavourable assessment.

Whether a belief is to be judged true or false has nothing to do with whether it has a cause. [Bloor (1976), page 14.]

It may seem unobjectionable, and not particularly relativistic, to separate questions of explanation and evaluation in this way. One might evaluate the beliefs of others as rational and true and think that since this could be a happy accident, one ought not to let that influence one's explanation. Equally, one might explain the beliefs of others as caused by factors usually associated with falsehood and irrationality but allow that they may yet happen to be, if not rational, at least true.

But the separation intended by Barnes and Bloor goes much deeper. This becomes clear when it is extended, as the postulate of reflexivity requires, to the sociologist's study of his own beliefs. Although he regards his beliefs as true and probably as rational, the sociologist is forced to deny that his regarding them so goes in any part to explain why he holds them; otherwise assessment would influence explanation. Equally, although he finds that his beliefs on a certain topic are fully explained by factors usually associated with falsehood and irrationality, he is not thereby given any reason to think again about them; otherwise explanation would affect assessment. Barnes (1974) embraces the result.

> A deterministic account of the creation of the arguments presented here is perfectly possible and acceptable. Even if the account cited 'external' social factors, this need not influence the evaluation of the knowledge thereby explained. [page 155. Compare Hesse (1980), pages 49—50.]

The reflexive case brings out the relativistic aspect of the claim made by Barnes and Bloor in their equivalence thesis. The idea is not just that methodologically it is advisable to keep explanation and evaluation apart. It is a metaphysical thesis to the effect that the causal factors in virtue of which beliefs are explained have nothing whatsoever to do with how they should be evaluated. Questions of evaluation — questions of truth and falsity, rationality and irrationality — float free of the relation between beliefs and the world encountered by believers. They are epiphenomenal matters on which believers make a judgment by the standards of the local context, without that judgment reflecting or initiating any relevant causal connection.

So much then for the relativism associated with the strong programme. In conclusion I want to emphasise that this particular form of relativism has a distinctively conservative cast. It means that no one ought to be troubled by having his beliefs sociologically explained, even explained in a manner that seems to debunk them. He can go on holding the beliefs in the face of any such explanation, for he need not think that they are any the less justified in the local terms of justification that he endorses.

Barnes gives nice expression to this conservatism.

> A scientific sub-culture, with its own esoteric procedures, competences, objectives and standards, is just like any other. Take painting, for example . . . If artists respond to a demand for altarpieces, or for prestige extravaganzas, and consequently modify their methods, sensibilities and standards of judgement, it is not assumed that by virtue of that very fact they have devalued their art. This is how things should stand also in the empirical study of science, and increasingly it is how they do stand. [Barnes (1982), page 117.]

The point of the sociology of knowledge is to understand beliefs, not to change them.

The conservatism supported by the strong programme gives it an element in common with the weak programme that it seeks to replace. If sociology is supposed to concern itself only with beliefs that are irrational, and presumably that are irrational by common consent, then equally no one in the ranks of regular believers, and certainly no one in the ranks of scientists, has anything to fear from the sociologist's probings. The weak programme makes it illicit to probe respectable thoughts; the strong ensures that the probing will be innocuous.

2. RELEASING THE STRONG PROGRAMME FROM RELATIVISM

To an outsider the most striking thing about the strong programme is that it can be quite naturally construed so that it does not entail the relativism with which it has been associated. The postulates quoted at the beginning may not be inconsistent with the conservative relativism which Barnes and Bloor read into them. But they are also consistent, it seems to me, with a critical stance which leaves open the question between relativism and its opposite.

Consider the following principle, which gives expression to a common notion of the ground for self-criticism: one should regard one's beliefs as rational, i.e., one should regard oneself as rational in holding them, if and only if one sees them as supported by appropriate as distinct from inappropriate causal factors — that is, by factors which tend to produce true beliefs. [See Bloor (1976), pages 32—39.] One should regard them as rational if one sees them as caused in this way, because rationality consists in their being produced by factors conducive to truth. One should regard them as rational only if one sees them so, because when one regards them as rational one must assume that one's regarding them in that way — a factor surely conducive to truth — helps sustain them.

Barnes and Bloor are committed by their conservatism to denying the truth of any principle of this kind. And yet it is striking that one can endorse the principle without making any commitment for or against relativism. One might still hold the relativistic thesis that different contexts, with different standards of appropriateness and rationality, may generate inconsistent beliefs and that there is nothing to make one context better than another. One might be non-relativist and deny that different contexts could produce inconsistent beliefs. Or one might be anti-relativist and admit that different contexts could do this but maintain that at most one such context employs the right standards of appropriateness. The principle is compatible with the full range of positions.

If we consider the strong programme under the assumption that the principle is sound, or even under the assumption that it may turn out to be sound, then a quite non-conservative reading suggests itself. Causality: the explanation of belief is always causal, whether or not it is the sort of explanation that can be comfortably endorsed by the believer, invoking features that he, given his context, thinks appropriate. Impartiality and symmetry: the sociologist ought to explore the social antecedents of every sort of belief with an open mind as to how it is to be explained. And reflexivity: the sociologist ought to be prepared to apply this approach to his own beliefs, continuing to maintain those beliefs only so far as there is a persuasive causal account of them that he can endorse.

Unlike the conservative construal offered by Barnes and Bloor, this reading of the strong programme gives it a critical cast. The idea is that the sociologist ought to approach his task, open to the possibility that what he discovers will be found subversive by those whose beliefs he explains. He is open to this possibility because, while he does not let his rational assessment of beliefs inhibit his attempt to explain them causally, he does admit that the causal explanation he comes up with may motivate a reconsideration of their rational status on the part of the believers; indeed it may also generate a reconsideration on his own part too. This construal does not protect the subjects of the sociologist's investigation from the effect of his work. It allows for the possibility of a critical influence.

The crucial difference between the two readings comes in the way in which the divide is drawn between explanation and evaluation. The conservative, relativistic reading asserts that neither should influence the other, because there is no connection between causal origin and rational or veridical status: this has the standing of a metaphysical thesis. The critical reading says that rational evaluation should not influence causal explanation, because to let it do so would be to introduce an uninformed prejudgment, but that causal explanation may be expected to affect rational evaluation, at least on the part of believers. This line is pressed on purely strategic grounds; there is no suggestion of a metaphysical thesis. [See Collins (1981) for inklings of a similar approach.]

I propose that the strong programme ought to be understood in the critical fashion rather than in that which presupposes a conservative relativism. I will mention three considerations in support of that proposal.

The first is that the critical reading is more sensitive to the beliefs of the subjects whom the sociologist of knowledge studies. It leaves it an open question as to whether or not something like the principle of self-criticism is sound, whereas the conservative construal has to assume that it is unsound. But the principle of self-criticism is implicit in ordinary practice, since when people take their beliefs seriously — when they see them, for example, as science rather than ideology — then they distinguish between debunking and non-debunking explanations of them and they feel required to respond to debunking ones. Those who support the conservative reading of the strong programme have to argue then that ordinary practice is misguided on this matter. Common sense has to be won around to a rather startling point of view. Viz.: "No sound basis has been found for a distinction between 'science' and 'ideology'." [Barnes (1982), page 111.]

A second consideration is closely related. It is that because of the metaphysical thesis required for its form of relativism, the conservative reading is philosophically more committed than the critical one. Barnes and others make great play of the alleged fact that unlike their opponents they do not shackle empirical work with philosophical presuppositions. But while they may do better in this regard than defenders of the weak programme, they do considerably worse than someone who construes the strong programme in a critical fashion: such a person can remain neutral, after all, on the issue of relativism. There is irony then in the following remark from Barnes (1979):

> It was not by following the philosophical orthodoxy on scientific rationality that the history of science made its enormous contribution over recent years, and generated such a wealth of philosophical problems; nor, at the present time, is there the slightest

indication that the field would benefit from tying itself to the apron-strings of philosophy. (page 253).

Indeed.

The third reason why I propose that the strong programme should be construed in the critical fashion is that it remains just as attractive under this reading as it does under the other. It does not have to deploy any controversial distinctions. It bears the stamp of a truly empirical, open-ended research programme. And it holds out the prospect of some iconoclastic results. Indeed it does this more surely than the other reading, for the prospect here is that some results may challenge believers to counter the explanation or reconsider their beliefs. The prospect there was only that some results would scandalise those who held a mistaken conception of the status of scientific belief.

Finally, a qualification. I hold that how utterances are interpreted depends on how they are explained, or at least on how utterances involving the same words are generally explained. [See Macdonald and Pettit (1981).] This means that, consistently with the claims of a scientific theory being construed in a certain way, not all those claims — I assume that some of the crucial terms occur only in such claims — can be explained in a manner which, from the point of view of explainer and explainee, is debunking. I therefore take the impartiality and symmetry postulates to entail, not that all beliefs on a given topic can be debunkingly explained — *salva interpretatione* — but that any particular claim or family of claims may be susceptible to such an explanation. [Compare Turner (1981).] It is true distributively of any specific claim that it may be explained in that way; but it is not true collectively of all claims that they may be so explained.

This is a qualification, because it takes from the strength of the strong programme. But notice that the reduction in strength does not come of imposing a critical rather than a conservative reading on the programme. If my case for the diminution is sound then it carries under either construal.

3. EXPLAINING THE RELATIVISM OF THE STRONG PROGRAMME

What we have seen is enough in my view to make us prefer the critical to the conservative reading of the strong programme. But it may be useful, in conclusion, to examine the likely pressures that drive Barnes, Bloor and the Edinburgh School in general towards the conservative construal. If it turns out that they are pressures which we need not feel then this will support us in our preference.

A first thought might be that they find relativism independently attractive and that this is what motivates the reading. But the thought leads nowhere, since the critical reading of the programme, as we have seen, is also compatible with relativism. Notice too that members of the Edinburgh School are generally anxious to suggest that they are driven by their approach to adopt relativism, not the other way around. [See Barnes (1974), Epilogue.]

Second thoughts raise the possibility that it is an independent commitment to conservatism rather than relativism which motivates the Edinburgh reading of the programme. If this hypothesis is sound, then conservatism must entail relativism and conservatism must look like an independent commitment which it is plausible to ascribe to members of the School. Both conditions are fulfilled and I espouse the hypothesis for that reason.

Conservatism, as it is understood here, holds that no believer ought to be disturbed by the causal explanation of his beliefs, no matter what the factors invoked. He ought not to be driven to reconsider their truth-value; he ought not to be driven even to question their rationality. Such a thesis requires the view endorsed by Barnes and Bloor, that matters of evaluation float free of the causal origins and connections of beliefs. And that view amounts to their rather distinctive brand of relativism. What will have to be said is that while the standards governing evaluation differ from context to context, no one set of standards engages any more than others with the causal bonds that link believers to their world; and, this being so, that no one context can be hailed as superior to others.

I conclude that an independent commitment to conservatism would readily explain why the Edinburgh School takes a relativistic stance on the strong programme. But is it plausible to ascribe such a commitment to them? I believe that it is.

The reason is that they see conservatism as the price that must be paid for a sort of value-freedom which they espouse: a value-freedom that they describe themselves as their naturalism.

Sociological accounts have no bearing upon whatever evaluations one may wish to put upon science; indeed, the major reason why such accounts are frequently self-described as 'naturalistic' is simply that they have no evaluative axe to grind. [Shapin (1982), page 187.]

The idea is that if the sociologist is to see his enterprise as value-free then he must disavow any intention to challenge or change those whom he studies.

A good expression of the idea is found in Barnes and Shapin (1979):

It might appear at first that what is being talked of here is a possible reversion to the state of affairs before the last world war when a significant movement of Marxists and radicals sought to expose science as a function of its social context. But this is not at all the case. In the 1930's both sides to the great debates accepted the importance of the internal/external dichotomy, and both sides recognised that sustaining their discourse were opposed methods of evaluating science and opposed policies towards it. In contrast, the current move to naturalism is not systematically related to any distinctive evaluation of science or policy towards it. Naturalism closes no evaluative or political options; it merely ejects them from historical practice. It may be, indeed, that the main impetus to naturalism stems from professional, disciplinary considerations on the part of historians and others engaged in the study of science. A recognition that explicit evaluative concerns and commitments are not conducive to good history may be the major factor. (page 10)

I conclude that not only would an independent commitment to conservatism explain the Edinburgh reading of the strong programme; it is also plausible to ascribe such a commitment to them. They take on that commitment with their fundamentalist renunciation of the influence of values.

The conclusion leaves us with one last question. Ought we to question our own reading of the strong programme, out of a wish to eschew values? I hold that even if that wish were reasonable, it does not provide a reason why we should raise a single query about the critical reading. The sociologist of knowledge who understands his work critically does not impose his values on it; in this regard he differs from the adherent of the weak programme. All he does is to endorse a conception of his work under which it may constitute a challenge to believers. Whether it does or not, he is prepared to leave to them. There is no reason here even for the most ardent evangelist of value-freedom to raise a protest.

Under the critical reading, the sociology of knowledge begins to look like the sort of discipline which critical theorists of the Frankfurt stamp might approve. It promises to offer a genetic form of the Ideologiekritik of science: a form of critique based on the genesis of the beliefs in question. [See Geuss (1981).] But lovers of value-freedom should not be disturbed, for one difference remains. This is that under my reading the sociologist of knowledge need not claim to know which of his findings will be challenging, which not. A fortiori then, he need not be moved to seek certain findings rather than others out of a belief — a value-laden belief — that those findings will be challenging or encouraging in their effect.[1]

Australian National University

NOTE

[1] This article derives ultimately from a paper read to a meeting in the University of Leeds, 1979 of the Theory Group, British Sociological Association. I am grateful for comments on matters in the paper to Robert Nola and David Papineau; I found Papineau (this volume) a useful stimulus to my own thought. The paper was written during an Overseas Fellowship in Churchill College, Cambridge and I am also grateful to the College for the facilities put at my disposal.

REFERENCES

Barnes, B. (1974) *Scientific Knowledge and Sociological Theory*, London: Routledge and Kegan Paul.

Barnes, B. (1979) 'Vicissitudes of Belief', *Social Studies of Science* **9**, 247—63.

Barnes, B. (1982) *T. S. Kuhn and Social Science*, New York: Columbia University Press.

Barnes, B. and Bloor, D. (1982) 'Relativism, Rationalism and the Sociology of Knowledge', in Hollis, M. and Lukes, S. (eds.), *Rationality and Relativism*, Oxford: Basil Blackwell, pp. 21—47.

Barnes, B. and Shapin, S. (1979) (eds.) *Natural Order*, London: Sage Publications.

Bloor, D. (1976) *Knowledge and Social Imagery*, London: Routledge and Kegan Paul.

Bloor, D. (1981) 'The Strengths of the Strong Programme', *Philosophy of the Social Sciences* **11**, 199—213.

Collins, H. M. (1981) 'What is TRASP? The Radical Programme as a Methodological Imperative', *Philosophy of the Social Sciences* **11**, 215—24.

Geuss, R. (1981) *The Idea of a Critical Theory*, Cambridge: The University Press.

Hesse, M. (1980) *Revolutions and Reconstructions in the Philosophy of Science*, Brighton: Harvester.

Laudan, L. (1977) *Progress and Its Problems*, London: Routledge and Kegan Paul.

Macdonald, G. and Pettit, P. (1981) *Semantics and Social Science*, London: Routledge and Kegan Paul.

Papineau, D. (this volume) 'Does the Sociology of Science Discredit Science?'

Shapin, S. (1982) 'History of Science and Its Sociological Reconstructions', *History of Science* **20**, 157—211.

Turner, S. (1981) 'Interpretative Charity, Durkheim, and the Strong Programme in the Sociology of Science', *Philosophy of the Social Sciences* **11**, 231—43.

KLUWER ACADEMIC PUBLISHERS GROUP
Book Review Department
P.O. Box 989, 3300 AZ Dordrecht, The Netherlands

1308

Dear Book Review Editor,

We are pleased to send you a review copy of our publication

Title : Relativism and Realism in Science

Author/Editor(s) : Nola

ISBN : 9027726477 Publication date : 23/06/88

Publisher : D.Reidel Publishing Company

Price : Dfl. 155.00 US$ 79.00 UK£ 48.00

We would appreciate receiving two clippings of your review in due course

K. HAHLWEG AND C. A. HOOKER

EVOLUTIONARY EPISTEMOLOGY AND RELATIVISM

1. THE MANY ROADS TO RELATIVISM

> Wide is the gate and broad the way which leads to destruction, and many there be that travel that way.

The relativist holds that in some crucial respect or other our cognitive abilities are simply too restricted or weak to determine our cognitive development. The essence of cognitive relativism is to deny that humans have access to methods for objectively criticising their own cognitive commitments; in particular, relativists argue the absence of critical methods which would transcend the conditions which produced those commitments. For example, it may be argued that theories are underdetermined by empirical experience and hence that there is no way to choose rationally among empirically equivalent theories and hence all knowledge claims are relative to the particular theoretical tradition within which they are made. Perhaps forces external to science itself may causally bring about a scientific revolution, but at least in the relativistic respects the revolution will not have been cognitively chosen, it will simply have been caused.

There have always been many different routes by which philosophic relativism can be reached. We begin by reviewing some of those which have proved most popular in contemporary analytic philosophy. We go on to argue that an *appropriately constructed* realistic evolutionary epistemology is anti-relativist. In what follows we shall for convenience restrict our attention to evolutionary epistemologies for science.

Argument 1

Either there is available a strongly normative philosophy of science independent of any appeal either to the content of science or the decisions of the scientific community, or there is not. If there is not, any attempt to justify a philosophy of science as normative must involve an appeal to what is genuine science, as opposed to the remainder which is

Robert Nola (ed.), Relativism and Realism in Science, 93—115.

either faulty science or non-science. But this appeal creates a vicious circularity since it is the business of a normative philosophy of science to delineate genuine science. Hence relativism follows. But there is no plausible philosophy of science which does not make essential appeal to either the content of science or the judgements of the scientific community. Hence relativism must be accepted. *Comment*: Both the logical empiricists and the critical rationalists (Popper and followers) accept some such argument as this and therefore seek to construct the strongly normative, objective philosophy of science which the first horn of the disjunction requires. Most relativist positions evidently presuppose this argument (often tacitly) as the beginning point for their particular considerations in favour of the second disjunct and some species of relativism held thereby to follow.

Argument 2

Either there is available a foundation for epistemology or there is not. (Here a foundation for knowledge is some category of judgements whose truth and objectivity *vis-à-vis* competing knowledge claims is strongly guaranteed in some way.) If there is no epistemic foundation then relativism follows since we have no way to objectively adjudicate competing knowledge claims. There is no objective epistemic foundation, hence relativism must be accepted. *Comment*: Empiricists are especially noted for this approach to epistemology and in their case empirical experience or observation is held to be the appropriate foundation. But in their own, opposite way the critical rationalists from Plato through Kant to Popper may equally be seen as attempting to provide a strongly (self-)guaranteeing foundation to knowledge. And contemporary critics of epistemic foundationalism such as Feyerabend have been quick to move from its rejection to various forms of relativism.

Argument 3

Either philosophy of science is able to justify a set of epistemic values or utilities as objectively correct (e.g. as applying to any rational cognitive enterprise) or it is not. If it is not then relativism follows since we must conclude that each cognitive enterprise is driven by its own values or interests. There is no way to justify a set of epistemic values or utilities as objectively correct. Therefore relativism. *Comment*: This argument has emerged out of continental hermeneutic and Marxist critiques of knowledge and also from within the sociology of knowledge tradition. Critical rationalists in the continental tradition (e.g. Habermas, likely Bhaskar) evidently accept it and try to meet its anti-relativist conditions.

Argument 4

Either a radically non-epistemic, correspondence theory of truth is justified or truth as warranted assertibility (or some weaker theory still) is the most that can be philosophically justified. Anti-relativism requires the correspondence theory of truth while warranted assertibility or weaker notions of truth lead to relativism. (This is so since what is warrantedly assertible will depend on the conditions prevailing within the community within which the assertion is warranted. To put it another way, coherence in various respects with the body of community beliefs dominates warranted assertibility and hence the very meaning of truth in the latter alternative.) But the correspondence theory of truth cannot be philosophically justified (perhaps cannot even be coherently stated). Therefore relativism. *Comment*: This argument has recently re-emerged among analytic philosophers of language (e.g. Dummett and Putnam). Alternatively, Ellis argues that the correspondence theory leads to scepticism (since humans could always be wrong) and hence to a warranted assertibility notion of truth and to relativism.

Argument 5

Either it is possible to demonstrate objectively for an epistemology that the use of its methods will lead science to converge on the truth, or it is not. It is not possible to give any such demonstration. But then relativism follows since we can have no objective reason to accept that science has made any progress towards the truth. *Comment*: This argument has been a recent focus of debate in U.S. philosophy of science (e.g. Boyd, Leplin). Its second premise has been supported by the God's Eye and Underdetermination arguments. The God's Eye argument is to the effect that there is no way for human beings to step outside of their conceptualisations in order to compare theories and reality directly, so theories can only be internally compared with other theories. The Underdetermination argument is this: theories are empirically underdetermined by observations and observations are the only objective criteria for judging theories, so choice among competing theories cannot be wholly objectively determined.

Argument 6

Sufficiently complex and/or deep theories are semantically incommensurable. There is no way to make a rational adjudication between semantically incommensurate theories, so knowledge is relative to theoretical commitment. *Comment*: An argument particularly promoted by early Kuhn and Feyerabend. There is an extended version in which theories are replaced by Lakatos' research programs.

Argument 7

Scientific commitments are importantly (perhaps even wholly) determined by pragmatic factors which operate quite independently of truth and falsity. So there is no objective cognitive way to choose among competing cognitive commitments. So relativism is true. *Comment*: This kind of argument is favoured by many sociologists of science and comparative anthropologists of culture (e.g. reality is taken as a social or cultural construct). A version of it is found in many continental hermeneutic and Marxist writings.

These are the general arguments for relativism found in contemporary analytic literature. They have tended to be accepted by both sides to the dispute, the relativists arguing that one or more premises are true or reasonably assertible, the anti-relativists attempting to show that they can meet the conditions required of them by the arguments. Typically, contemporary philosophers string together several of these arguments into a single attack on anti-relativism. Much of the relativists leanings detected in Kuhn by various critics centre around elements of Argument 1 (insistence that philosophy of science must learn from the history of science), 2 (assertion that facts are theory-laden constructs) and some version of 5, 6 or 7 (the claim that canons of theory choice are themselves internal to scientific communities). Feyerabend is well known for his advocacy of arguments 1 (history shows that 'anything goes' in scientific method), 2 (observation is theory-laden), 3 ('What is so great about truth?') and 6 (the meaning of a theoretical term is a function of its theoretical role — the doctrine of implicit definition), with smatterings of most of the other arguments thrown in.

Contemporary analytic literature in philosophy of science provides us with an extensive and rapidly growing catalogue of people who have adopted one or more of these arguments and used them to attack critical rationalism and realism especially and to argue for various forms of relativism. (Kuhn and Feyerabend had already begun the relativist-based attack on empiricism two decades ago; at the present time that attack tends to be assumed rather than repeated.) For reference to this literature and to all those philosophers mentioned above, from a realist anti-relativist perspective, the reader is referred to Devitt (1984), Hooker (1986a) and Trigg (1980).

It is not our purpose in this paper to respond to these arguments in general on behalf of realism. [For this response see Hooker (1986).] Rather, we want to concentrate on the bearing of relativism on the construction and defense of a realistic evolutionary epistemology. To this end we focus our attention now on biologically based approaches to cognition.

It is not hard to see why philosophers have found it easy to believe that any seriously biologically based epistemology is bound to finish in relativism. Any such position will want to adopt an evolutionary approach and according to an evolutionary approach cognition begins from profound ignorance, ignorance not only of the nature of the world but even of what knowing itself is and how it proceeds. Not only science, but philosophy itself must be fallibilistically construed. For a thoroughgoing evolutionary epistemology, both our cognitive capacities and the cognitive commitments to which those capacities lead must have emerged through some process of variation (broadly 'trial and error'), selection (e.g. judgments concerned with accuracy, practicability) and retention. But in this case it is hard to see how there could be available a strongly normative philosophy of science independent of any appeal either to the content of science or the decisions of the scientific community (cf. Argument 1). It is hard to see how there could be a privileged epistemic foundation for knowledge, since all knowledge claims are fallible (cf. Argument 2). It is hard to see how a fallible philosophy of science would be able to justify a set of epistemic values or utilities as objectively correct when a central part of our history of learning is to learn about the nature and range of epistemic values which may be operative in intelligent systems (cf. Argument 3). And surely the only epistemic criteria for selection to which we fallible, ignorant creatures have access are criteria of coherence with the totality of our species experience to date (cf. Argument 4). Surely, too, both the God's Eye and Underdetermination arguments apply (cf. Argument 5). Since neo-Darwinists, like Marxists, want to deny to cognition independence of genetics (respectively economics) there will be a strong pressure to adopt a socio-biological version of Argument 7. The sweep of arguments would be completed if an evolutionary account of language found some version of Feyerabend's implicit definition plausible, as many would (cf. Argument 6). In sum, it looks as if anyone wishing to adopt an evolutionary approach to cognition will encounter the full sweep of relativist arguments and be drawn by each of them into a relativist conclusion.

Moreover, we note for later reference two further relativist arguments which begin explicitly from an evolutionary basis.

Argument 8

There is no justifiable notion of progress constructible within Darwinian adaptation, rather there is simply a shifting set of adaptations tracking (more or less imperfectly) the shifting ecological structure of the planet. For an evolutionary epistemology adaptation corresponds to empirical adequacy, a theory is well adapted when it is able to accurately predict and explain the features of its domain (environment) of application. Hence there is no justifiable notion of theoretical progress constructible in an evolutionary account of science. Rather there is perhaps simply a shifting theoretical structure which tracks (more or less imperfectly) the shifting socio-cultural-cum-technological environments we happen to devise. Or as van Fraassen (1980) recently put it: The theories we currently have, we have because they are the ones which happened to survive historically, not because they are in some global sense objectively the best.

Argument 9

According to an evolutionary account each species will develop a set of adaptations which is idiosyncratic to the initial genetic heritage of the species and the sequence of environments through which it has evolved.

Thus we can expect of any species which develops cognitive capacities among its adaptations that these capacities will be relatively idiosyncratic to the species concerned. Since the actual cognitive commitments of that species will be a function of its cognitive capacities, and quite possibly a function of many of its non-cognitive capacities as well, we must expect that the resulting cognitive commitments of a species will be relatively idiosyncratic to it. In which case we have a deep-going species relativism for knowledge; we should not expect, upon meeting an alien species, that we will necessarily share any significant cognitive content, framework or methods with it. This argument is developed extensively by Munevar (1981).

We are realists and we are anti-relativist. We also wish to take seriously a thoroughgoing evolutionary epistemology. It is our conviction that relativism represents a profound misunderstanding of the nature of cognition (for any species).

Relativism is a comfortable position. A relativist doesn't need to worry about being wrong — for there is no sense to being wrong in the relativistic respects. A relativist can remove the tensions from cognitive life. There is e.g. no need to worry about holding philosophy as both fallible and yet as providing a normative critique for the development of cognitive commitment; that development becomes instead simply a part of a scientific story, either a purely causal story or an anthropological story about human conventions. But we believe that risk taking and cognitive tension stand at the very centre of the nature of cognition, that cognitive dynamics cannot be understood in any insightful way without appealing centrally to these notions [cf. Hooker (1982), Hahlweg (1983)]. When a species develops cognitive capacities through an evolutionary process it risks its survival on the exercise of capacities of unknown depth and breadth to form cognitive commitments of unknown accuracy. It must do this both at the level of descriptive commitments and at the level of epistemic methods. The methods themselves then must both be construed as fallible conjectures and as the vehicles of normative critique of descriptions (theories). Not to understand this is to miss two of the central features of an evolutionary understanding of cognition.

For sheer plausibility, most relativists concede that there has been practical, technological progress brought about by science. (They show this practically by flying to conferences rather than walking.) At the same time relativists are committed to denying any deeper sense of progress to human knowledge. We hold that an adequate evolutionary understanding of cognition must find practical and cognitive progress inextricably linked. We shall exhibit the essence of our view in Section 3 below. Once again, we shall conclude, the relativist misses the very essence of evolving cognition. But this will not be before we have radically recast traditional approaches to evolutionary epistemology. We proceed first to its realist foundations.

2. EVOLUTIONARY NATURALIST REALISM

> But small the gate and narrow is the way which
> leads to salvation, and few there be that find it.

Paradoxically (perhaps) a good place to begin the construction of a defensible realism is with the acknowledgement of all those features of an evolutionary approach to knowledge which have just been claimed to support relativism! Realism should be nothing if not realistic. If we are indeed a species after the manner which contemporary biology teaches, then it is necessary to take on board those features of cognition which follow plausibly from such an approach. But we insist emphatically that these features do not support relativism. They may argue the abandoning of traditional philosophies of science, empiricisms, critical rationalisms and even traditional realisms, but in our view they ultimately lay the foundation for a much more penetrating, more critical anti-relativist realism.

Both of us have written at some length about realism from an evolutionary perspective elsewhere [Hahlweg (1983), Hooker (1975a), (1982) and especially (1986a)]. Rather than repeat that material here, we want to go on to develop a central feature of our current conception of evolutionary epistemology and indicate how it contributes to the anti-relativist arguments already advanced in the foregoing writings. In this section we confine ourselves to some very brief remarks about countering general anti-relativist arguments.

The essence of defeating the relativist arguments is to take absolutely seriously the idea that there is no infallible or First philosophy, that all philosophy is fallible and must be learned as we go along. From this it follows immediately that there cannot be a sharp separation in kind between normative and descriptive; indeed that philosophical theories are just that, theories about the norms of intelligence in exactly the same way that scientific theories are theories about other structures in the world. More than a decade ago, C.A.H. challenged philosophers and scientists to specify in what the normativeness of philosophical theories consisted that was over and above the ways in which scientific theories may be used to specify the kinds of experimental procedures which are relevant, estimate the strength of evidence and criticise the validity of data [Hooker (1975a), cf. (1982)]. Why can we not be exploring normative theories simultaneously with exploring scientific theories? Of course, this makes the whole enterprise riskier — it is always possible that we may run ourselves into a methodological dead end. But life is a risky business, the essence of being intelligent is not to eliminate risk — Popper has already taught us that this kind of caution is an evolutionary dead end — the essence of being intelligent is to take risks which are productive of new knowledge and yet (*ceteris paribus*) safe-fail (tolerable in failure). Indeed, we regard the real key to understanding scientific development to lie, not in the area of theoretical development (where all conventional philosophies of science look), but instead in the area of methodological development. And we argue in Section 3 below that this provides the right kind of connection to evolutionary biology as well.

The essence of our realism is the fallibilistic exploration of conceptualised proposals at all levels from perceptions through theories to methodologies, metamethodologies and so on 'up'. Each level acts as a normative guide and critic of the level 'below' it and the level 'below' acts as data which the higher level is required to explain and understand. But in this dynamics a change at any one level may, and quite often does, ramify across several levels (consider e.g. the multi-level impact of quantisation). All levels are in interaction with all others. Moreover, all levels are in interaction with the non-cognitive world twice over, via technology and via the social structure of scientific activities. Technologies are conceived and designed through theory and experience and they in turn extend our experience and also transform the societal conditions in which scientific activities occur. Indeed, technologies have not only extended and corrected our senses, transforming our interaction with the world, they may someday transform our cognitive structures, our brains, themselves. And the whole modern international apparatus of science would be quite impossible without the economic and communicational transformation wrought through contemporary technologies. (We may even succeed in undermining the whole scientific enterprise by destroying its biological basis, because our physical technologies have outrun our institutional and psychological technologies.) The upshot then is that science is a highly dynamic cognitive system in which all the components are involved in an open-ended development which is both transforming every cognitive level to the system and simultaneously transforming the environment in which the cognition is developing. Just this is of the essence of the biological role of cognition (see Section 3).

And this is a thoroughly realist position in the classical sense of the term. It is precisely the existence of a real world external to the mere disposition of our current beliefs that both makes this kind of open-ended cognitive dynamic necessary and which anchors the sense of it as a genuine exploration rather than simply a convoluted invention. In this exploration truth must play a substantive and complex role as a cognitive ideal, never certainly and never fully attainable, but giving global sense to the cognitive process. Indeed, it is only when truth is reduced from this role to a narrow semantic property of language that the contemporary arguments for abandoning the correspondence theory of truth takes any hold [cf. Hooker (1986b), Section 3.1 and Argument 4 above]. It is only from a realist perspective that we are able to deal effectively with scepticism as the confusion of risk with illusion. The essence of the sceptical argument is the move from the demonstration that error is possible according to some epistemology to the claim that said epistemology can provide no valid insight into the nature of knowing. But if there is a real world not tied but weakly, and then only causally (not logically), to human cognitive processes then it is of the essence of our ability to learn about intelligent cognition that we be able to fallibly explore alternative cognitive methodologies and alternative cognitive styles more generally [cf. Churchman (1968), (1972); Hahlweg (1983)].

From this point of view Arguments 1, 5, 8 and 9 are all non sequiturs, while Arguments 2, 3, 4, 6 and 7 contain false premises. The preceding discussion will already have given some indication of how we would argue in these instances. [See again also Devitt (1984) and Trigg (1980)]. Rather we shall turn forthwith to our notion of methodological dynamics *vis-à-vis* evolutionary dynamics and indicate its relationship to relativist concerns with objective adjudication among epistemic claims and traditions, addressing directly Arguments 2 and 5, but especially 8 and 9, and indirectly Arguments 3, 6 and 7.

3. TOWARDS A DYNAMICS FOR EVOLUTIONARY EPISTEMOLOGY

Ever since the advance of modern science in the 17th century, its development has been viewed as progressive. Bacon even argued that progress in science is equivalent to societal progress and that "the sure march of science" would eventually result in a general betterment of mankind. The times when scientific progress was seen as being equivalent with societal progress are certainly over. In the light of nuclear weapons and ecological disasters, it has become clear that scientific progress does not automatically yield societal progress, and that science has not only to be understood as a problem solving but also as a problem creating activity. But this latter is true also within science, as Popper has emphasised; it is the relation of the two which must be understood.

Recall that it is typical of relativists to concede that technological progress has taken place (and as a causal result of science, at least in this century). Relativists may dis-value that progress, but that is another issue. It is simply not plausible to assert that our capacity to exercise (limited, but real) technological control in an ever widening variety of circumstances is illusory. But we shall now argue that this kind of progress with technological control is intrinsically associated with cognitive progress in understanding the world, the very kind of progress the relativist wishes to reject. In short, the relativist cannot both accept technical progress and deny cognitive progress. To do so is to misunderstand the central nature and dynamics of cognition.

In the course of evolution organisms increasingly learn how to cope with the variety of environments within which they find themselves. In adapting to a new environment they also transform it. In the same vein a scientist learning about the world also transforms it. The interpenetration between living things and their surroundings leads to the creation of an increasingly diverse world. Likewise, science in learning from the world also changes it, a dialectical process which results in the creation of ever new theories, methods and environments. As a result of this interaction we find ourselves progressing into ever new areas of reality, understanding and transforming it as we go along. We do not know where the future will lead us, we do not know which fundamental restructuring of our conceptualizations may be necessary. But we do know that we are making progress because we can operate successfully in areas of reality of whose existence our forefathers did not even dream. The scientists of the past did not plan modern science any more than natural selection planned the emergence of modern species. It is the inbuilt dynamic of the evolutionary process which pushes us across new frontiers into realms where no other species on earth ever penetrated.

It is highly controversial to speak of 'progress in biological evolution'. Such talk seems to imply that there are goals to be achieved by the evolutionary process, that the evolution of all species is directed toward that goal, a view prevalent in pre-Darwinian thought but not tenable any more. Its untenability is behind relativist argument 8 above. But this is the wrong focus for progress, in both biology and cognition.

If the concept of progress has been accepted at all, then it is in terms of 'progress in adaptation', implying that a species ill-adapted to a certain environment will in the course of evolution develop those physical and/or behavioural characteristics which will make it better adapted to the environment in question. Even this conception is problematic; but while we may allow that it makes sense to talk about progress in adaptation to a particular environment, it makes no sense to speak about progress in adaptation as such. If longevity is any judge, species were very well adapted in the past to their environments. Species persisted for many millions of years; if they became extinct then it was because that environment changed, an uncertain future which may be shared by the human species as well (and likely in less time).

Still, this does not mean that every concept of 'progress in evolution' is necessarily either meaningless or anthropocentric. It is of course true that the concept of progress is intrinsically comparative: it implies that a change occurs and that the end result ranks higher on some scale than do earlier conditions. For example, a biologist investigating the evolution of bird flight may point to the development of strong wing muscles or an extremely efficient blood circulatory system as contributing to the animal's flying capabilities. The stage of evolution when these characteristics appeared first therefore constitutes progress with respect to the bird's ability to fly. It does not imply, of course, that evolution intended to create birds with this capacity. The term 'progress' taken this way implies neither anthropocentrism nor that there is something called 'progress' other than that to be specified with respect to a certain organic or behavioural capability. Thus, we can use the notion 'progress in evolution' as long as we specify which organic or behavioural capability has improved in its course.

In the present context we are interested in 'cognitive progress'. It is beyond dispute that during evolution species with increased cognitive capacities evolved. Not all species get 'smarter', but some do. The question to be asked is: What are the conditions which foster the improvement of these capabilities? We want to know which conditions favour the development of more sophisticated cognitive abilities in contradistinction to those which tend to favour the retention of a more primitive cognitive organization. Once we understand the processes which are instrumental in improving a species' cognitive abilities we may ask if a similar model could be used to explain the evolution of science.

We emphasize that the environment is of utmost importance for the understanding of evolutionary change. Some creatures live in very homogeneous environments, that is, a comparatively small number of forces impinge upon them. Different creatures, by contrast, may experience a multitude of environmental forces, i.e. a heterogeneous environment, even within the course of one day. For example, the temperature in a forest varies substantially from the cool shadowy areas underneath big trees to the hotter open areas. Some animals may always stay in the shade, confining themselves to a fairly homogeneous temperature environment, while others, moving from the cool forest floor into the sunshine of a meadow, experience a more heterogeneous temperature environment. Species which survive successfully in a heterogeneous environment are not merely adapted to a specific environment, they must be *adaptable* to a variety of environments. It was emphasised by the distinguished embryologist and evolutionary theorist, C. H. Waddington, that the property of adaptability is of utmost importance if we want to understand the evolutionary process [Waddington (1975), cf. Hahlweg (1981)].

During periods of environmental stability species maintain a state of equilibrium with their environment. Any change in the genetic makeup of a species is likely to be harmful and will be eliminated by selective forces. Once, however, the environment begins to change the organisms have to adapt to the new conditions. In general it will not be those species which are well adapted to their current environment which will do best in the dynamic conditions. Rather, those which are adapted to many different environments will have an edge in the evolutionary game. They appear to be 'preadapted', but in fact they are only adaptable. For example, a species which has been coping with wet as well as with dry conditions is more likely to survive a climate change altering humidity levels than is a species which can thrive only in a narrow humidity range. In particular, in case of the environment changing very fast, a species cannot simply wait for the 'right' mutation to turn up; it will become extinct unless it is already adaptable to a variety of environments. Thus, frequent or fluctuating changes of the environment will favour the development of species which are not only adapted but also adaptable. On the other hand, in a stable environment adaptability is of little use and will therefore not be favoured by natural selection.

There are obvious limitations to the range of environments to which an organism can adapt. With intelligence, however, organisms can escape adverse circumstances by detecting and moving to the best living conditions or by modifying their environment to suit their requirements. Thus a highly effective way of improving adaptability is to improve cognitive capacities. (There may be other ways to obtain adaptability, e.g. by regressing to a simpler chemistry, cf. cockroaches *vis-à-vis* humans, but we don't pursue these alternatives here.) The human species is by far the most adaptable of all species. Neolithic man and woman already learned how to cope with the most adverse circumstances. In the cold of the ice ages they made use of fur and fire, in the heat of the desert they learned to preserve water and search for shelter from the burning sun. Their hunting and gathering capacities let them survive in extremely barren country. This outstanding case of adaptability was not due to a particular physical prowess — many animals surpass us in this respect — but due to the high intelligence and sociability of the early humans.

Adaptable species do not only passively adapt to certain environments, they also change them and thereby create new environments. Wombats dig burrows, beavers build dams. In fact, the evolutionary process has transformed our planet. For example, forests, savannahs and tundras were all created by evolutionary processes. Even the oxygen in our atmosphere is said to be a result of biological processes. Those species which are adapted, not to a narrow ecological niche but to a wider spectrum of environments, are also more likely to migrate and live successfully in new lands, transforming them in turn. In a dynamic interaction species change the world which in turn poses new demands on their adaptive capacities. Those who are too slow will perish, those who adapt fast enough will further increase the rate of change. This exponential growth is characteristic of evolutionary processes. Many species will earlier or later lose in the evolutionary gamble, their adaptability will not suffice. The alarming rate of species going extinct in this century provides ample evidence of the threats a continuously changing environment poses to its inhabitants. And this is the key to the notion of adaptability, and hence intelligence: highly adaptable species achieve their largest advantage in a dynamic environment. It is characteristic of adaptable species to destabilise their environment.

We follow Waddington and equate the notion of 'progress in evolution' with 'progress in adaptability' [cf. Hahlweg, (1981), 1983)]. This is an acceptable comparative notion of progress, yielding a partial ordering on species with respect to a class of environments. Note, however, what a double-edged notion this concept of evolutionary progress really is. Intrinsic to this notion is the idea that environmental change is absolutely necessary if progress is to occur. The same change, however, will be detrimental for many species, perhaps even for the initiating species.

Destabilising the environment is most evident in the human species. It is we humans changing natural habitats that is responsible for the unprecedentedly high rate of species extinctions this century. And it is our genetic technologies which are responsible for creating (domestic) species at an unprecedentedly high rate. (Nonetheless, our current over-all impact is evidently toward degeneracy, both genetically and ecologically — we are a very young and ignorant species.) Humans are, in fact, rapidly transforming the planet into a human artifact; water systems, plant and animal species, urban and industrial environments, even the global climate pattern are being transformed by human intervention and design. And this applies increasingly to humans themselves, both socially and psychologically [see Hooker (1982)]. Thereby follows a profound transformation in human decision making, from reaction to anticipation, from natural activity to deliberate design, from intuition to theory mediation. This transformation has deep implications for public policy making, education and ethics [cf. Hooker (1982), (1983)], but we don't pursue these here.

The 'vehicle' of adaptability and environmental change in our case is cognition and its 'engine' is science/technology. We hold that the scientific-technological process is a continuation of the biological process of adaptability dynamics. In a seemingly endless cycle we are confronted with problems which we solve but whereby the very act of problem solving creates new problems. This dialectical process drives the 'cognitive engine' of modern science. It is indeed the very dynamics discussed in Section 2 above. If we want to understand scientific progress we have to understand the dynamics of problem creating through problem solving. We also have to understand why in the course of this process not only the scope of science widens, but equally its depth as evident in the increasing power of scientific conceptualizations. To underpin this anti-relativist conclusion we examine further the transfer of biological concepts to their appropriate scientific counterparts and thereafter look into the dynamics of science in greater detail.

That local environment, or part of the total environment, to which a population of organisms is adapted is called its niche. Equally, a niche consists of all those forces impinging on a species with which the species can cope, i.e., with which its adaptations suffice to deal. We call that part of the cosmic environment which humans can conceptualize our cognitive niche. During evolution more adaptable species develop, their niches expand as evolution goes on. Likewise during conceptual evolution wc succeed in capturing an ever increasing number of parameters. Our cognitive niche grows as our capacity to explain, predict, and control grows.

The classic case is that of mechanics. Newtonian science yielded accurate explanation and control for a vastly wider range of conditions than did the Aristotelian tradition which preceded it, and two centuries of relative theoretical stability ensued. But when mechanics expanded its parameter ranges to the very small, very fast and very large, its conceptualisations broke down and adaptation required new theories (relativistic and quantum mechanics). Ultimately of more importance, these adaptations required new methods, mathematical, theoretical and experimental-technological. And the new methods have set in train the development of much more flexible, powerful 'higher order' conceptions of what mechanics is than was every conceived in the Newtonian scheme. As a result we do not merely have several new mechanics adaptations suitable for new parameter regimes, but also a new highly adaptable conception of the science of mechanics.

As for biology, so too in cognition the role of environmental heterogeneity and de-stabilisation is critical. As long as we are restricted to a comparatively homogeneous cognitive niche, the concepts we employ are not seriously challenged. Over many generations stability prevails, new concepts will be aborted in favour of old ones which have proven their usefulness (Kuhn's normal science).

Not only science but cultures generally exemplify this dynamics. Primitive cultures are examples of conceptual stability. Over very long periods of time they inhabit the same kind of environment. Their beliefs have proven to be sufficient for the mastering of their surroundings. They didn't feel any 'pulls and pushes' forcing them into new environments. They have been in equilibrium with their environment, both physically and conceptually. Once the environment does change, however, they often cannot adapt. Their belief system, as well fitted as it may have been to their traditional surroundings, proves insufficient in view of the new developments. They were adapted but not adaptable. Sadly, their inability to conceptualize new events often leads to their extinction as a separate culture (and we lose valuable insight into experiments in being human).

Western culture is the epitome of a change-oriented, environment de-stabilising culture, its advantage over other cultures lying in its science-based adaptability. Unsurprisingly, our time has witnessed the vanishing of primitive cultures at an unprecedented rate. These were people who often had survived a multitude of hostile physical environments, they were certainly adaptable in this respect. But their model of the world could not accommodate the changes brought by the white man. Their cognitive niche was narrow, they may have been cognitively adapted to their old environment, but they were not adaptable to the new one. A conceptual framework which does not only let humans be adapted to a certain surrounding but lets them rather be adaptable to many different environments, is of tremendous advantage at times when we are continuously encountering new changes. (Note that this is compatible with charging western culture with the neglect of many areas of personal and social experience, areas where we may profitably learn from other cultures. In a fully explored history, such defects will eventually show up as limitations on adaptability.)

The niche inhabited by western cultures has widened to a large degree during the past three centuries. We have expanded into many new environments on earth and are now even exploring outer space. This achievement is due to a technological development which in turn rests on concepts developed in science, in turn stimulated by technologically transformed conditions both internal (laboratory) and external to science. Our cognitive niche is directly correlated to our ecological niche. In order to understand cognitive processes we have to understand technological progress, in order to understand technological progress we have in turn to understand cognitive progress. The relativist cannot dissociate the two without foregoing insight.

Traditionally the evolution of science has been compared with biological evolution on the concept of adaptedness. If we were to take this conception as a model for science we would have to conclude that conceptualizations of primitive tribes have equal right as contenders for truth as those of modern science. Hence the relativism (cf. Argument 8). This, however, is the wrong model for scientific development. An understanding of progressive change in both biological and conceptual dynamics has to focus on adaptability, and for adaptability the interrelationship between a species and its environment is critical. [That this relationship is a dialectical one has also been emphasised by Lewontin (1978) and (1982)]. Hence for cognition an understanding of cognitive adaptability is at the focus of cognitive progress and to understand progress in cognitive adaptability we have to focus on the science-technology dialectical interrelationship as briefly sketched in Section 2.

Science claims to deal with the world in its wholeness. Scientists attempt to find what is invariant in nature as a way to exhibit objective structure. They need therefore scientific theories and practices which are not only suitable for dealing with a limited range of phenomena but with the vast variety of changes we observe in the world. Each conceptual niche constitutes a model of the universe but some niches are very narrow whereas others include a tremendous range of phenomena in their scope. To increase the breadth of our cognitive niche means to increase our access to reality, to transform our conceptions in such a way that they can accommodate a wider range of phenomena. A species' adaptations may prove insufficient when encountering a changing environment. It faces the challenge by evolving new adaptations. Sometimes a minor modification of the original physiological or behavioural equipment may suffice, at other times a restructuring of vital characteristics may be demanded. Adaptability is reinforced since it is the adaptable species that cope best with the changes. Likewise, a conceptual framework may prove insufficient when confronted with new phenomena. Sometimes a minor modification of existing concepts may suffice, at other times a complete restructuring of basic metaphysical presuppositions becomes necessary. Those methods or methodologies which provide for greater conceptual adaptability are thereby reinforced since it will be cognitive structures (in this case sciences) with highly adaptable methodologies which cope best with exploration and change. The new concepts are tested in the new environments holistically and practically, in the same sense as organisms are tested.

In praxis this means that testing is constructing, constructing experiments, applying technologies to social conditions, applying methodologies to the construction of new theories, scientific interrelationships, even new scientific institutions [see Hooker (1982)]. If the concepts employed suffice to create stable experimental conditions under which the test in question can be performed, then these concepts have proven to be adaptable to the new situation. If, however, stable experimental control is not possible, then the underlying framework of concepts may need to be revised. This need can arise anywhere, even from the most prosaic laboratory work.

For example, a scientist begins with a definite problem, say the investigation of the deposition of crystals of a specific kind out of the gaseous state. Using current theories, the scientist considers which parameters may be of importance, which situations may disturb and which may favour the deposition of regular, single crystalline structures. Following theory, an aparatus is constructed suitable for controlling the flow of gaseous substances, their temperature, pressure and chemical constitution, and experimentation begins. Eventually there may be success in creating the desired crystals under controlled conditions. In that case, the equipment will not have presented any challenge to the theoretical concepts employed in its design. But now assume instead that no crystallisation occurs. Instead an unknown liquid accumulates in the vessel which after analysis turns out to be a new, hitherto unknown compound. Let us further assume that according to current chemical knowledge it is thought to be impossible to synthesize such a compound. Apparently the specific experimental conditions were just right to create the new "impossible" substance. The scientist will have to search for a theoretical explanation in terms of current theory but such an explanation may not be forthcoming. In that case it will be necessary to postulate that one of the underlying theories employed in designing the experiment was at fault. After publication of the exciting experimental findings and ensuing conceptual exploration, it may be proposed that Molecular Orbital (MO) theory has to be amended. How do we test this kind of hypothesis? By constructing new 'impossible' compounds, i.e. compounds whose existence would have been denied by the previous MO theory, but are to be expected according to the new hypothesis [cf. Hacking (1983), see Hooker (1986b) Section 4]. Let us assume the hypothesis is confirmed. A major theoretical revision of the theory of chemical combination has taken place. The revised theory will from now on be employed in every laboratory. It will be used for reactions which differ in every constituent, in temperature, pressure, perhaps even in state. Very diverse compounds will be constructed and, of course, sometimes in the future the theory may encounter a new challenge and need to be amended again. More importantly for this history then, chemical methods, embedded deep into the foundations of chemical theory, will be broadened by the comparison of the two situations, old and new MO chemistry, and their future adaptability increased. Scientists will assert, quite properly, that deeper insight has been gained into the structure of substances.

As historical entities we begin by living in very narrow niches. Our conceptualizations of the world are necessarily equally narrow. We won't see farther than we are forced to do. Fact and fiction, myth and observation are intimately interwined. There is no stepping out of the niche into a god-like realm, cognitively or biologically. But once we are forced into new realms of reality our adaptations or conceptualizations may prove to be insufficient. According to our best guesses, our metatheories, our metaphysics and methods, we amend our received theories. Perhaps our framework is intrinsically too weak, perhaps we are so wrong that we can't make it in the new, changed environment. We may become extinct, either biologically, as a culture, or as theory-believers. But in some cases we are able to revamp our concepts and methods in such a way that they can deal with the new situation. In that case our cognitive niche has been expanded. Becoming more adaptable, we bravely march forward into new realms of reality — and we get defeated again. Earlier or later we find that our constructions won't suffice and cannot further carry the burden of leading us into new and strange lands. We have to question ever more deeply ingrained beliefs. Each change leads to new concepts, new constructions and new methods which, if successful, lead us into new environments. By being forced to reflect on even our deepest metaphysical presuppositions we continuously improve our cognitive structures.

The history of science provides ample evidence of more or less fundamental revamping of theoretical concepts and methods. Recent history has seen the questioning of very profound presuppositions such as the concepts of mass and energy, of space and time. Even our logic has been questioned by philosophers and physicists dealing with the problems of quantum mechanics. Our conceptual niche has been expanding continuously. New, unknown environmental forces are being encountered in the course of scientific research. If the concepts and methods at our disposal are adaptable they will accommodate the new factors. If, however, these are defective, they need to be amended. The amendation in turn is tested by construction. (We cannot test an individual scientific concept any more than we can test an individual physical or behavioural adaptation. As already Pierre Duhem pointed out, there are always many theories involved in every single experiment; [cf. also Hooker (1975a)]. If our controls fail we have to try again. If they succeed we will keep them until they, too, may need to be amended.

There have been many dead ends in evolution. Even species adaptable to many environments become extinct eventually. The demands of the ever-changing environments were too great, the basic structure of the organism not capable of undergoing the required changes. Likewise many, if not most, conceptualizations of the past became extinct. Their conceptual frameworks could not carry the heavy load of new discoveries. Alchemy and astrology may be cases in point. (Yet at the same time they may provide us with some valuable ideas for later use, just as other cultures surely can — a rich and widely shared genetic basis is essential for future adaptability.) Likewise we never know if the framework we are employing at present will be strong enough to support changes required in the future. Still, the fact that our concepts are fallible does not mean that one conceptualization is as good as any other.

The relativist wants to dissociate practical and cognitive progress, but for evolving creatures this isn't practicable. Relativist arguments that assume an adaptation-based epistemology are also working with the wrong conception of cognition (and evolutionary biology) — Argument 8 fails. And, *pace* Argument 9, while it is true that a species' knowledge will be partial, and idiosyncratically so, no relativist thesis follows. To the contrary, were we to meet an intelligent alien species, it would be an exercise of intelligence for both species to set about trying to fairly master the other's environment(s). There would then be a direct practical test of the merits of each species' system of cognition:

who can cope with the widest range of environments? There may be practical reasons why the test is not seriously conducted (e.g. we may declare war on them) and it may only achieve partial resolution for a long time, but none of this undermines the objectivity and realism of what is happening. Where there is overlap of mastered environments, there will be construction of relatable invariants and comparability of knowledge across the range of the weaker epistemic system. The cognitive exploration of reality is a risky, surprising and ever-incomplete adventure, but there are no good grounds for thinking it a relativist one.

4. CONCLUSION

The debate between relativists and realists is usually carried on in abstraction from any detailed considerations of cognitive dynamics, much less from biological models of those dynamics. We have begun this paper in the same vein. But we believe that this approach is basically mistaken. Fallibilism and the needed cognitive dynamics which follow are of the essence even of a defensible general realism. This is the lesson on which Section 2 focuses. In Section 3 we briefly develop a conception of cognitive dynamics which supports the kind of anti-relativist realism we suppose defensible and which is intimately related to a conception of evolutionary theory.

This is not the only dimension that needs to be developed for an adequate evolutionary epistemology, but the notion of adaptability dynamics stands at its core. It should be apparent that this conception of evolutionary epistemology departs substantially from the current candidates in the field and we believe only a theory along our lines has a future.

University of Newcastle

REFERENCES

Churchman, C. W. (1968) *Challenge to Reason*, New York: McGraw-Hill.
Churchman, C. W. (1972) *The Design of Enquiring Systems*, New York: Basic Books.
Devitt, M. (1984) *Realism and Truth*, Oxford: Blackwell, 1984.
Hacking, I. (1983) *Representing and Intervening*, Cambridge: The University Press.
Hahlweg, K. (1981) 'Progress Through Evolution? An Inquiry into the Thought of C. H. Waddington', *Acta Biotheoretica* **30** (1981), 103–120.

Hahlweg, K. (1983) 'The Evolution of Science: A Systems Approach', Ph.D. thesis, University of Western Ontario.
Hooker, C. A. (1975a) 'Global Theories', *Philosophy of Science* **42** (1975), 152–179.
Hooker, C. A. (1975b) 'Systematic Philosophy and Meta-Philosophy of Science: Empiricism, Popperianism and Realism', *Synthese* **32** (1975), 177–231.
Hooker, C. A. (1982) 'Understanding and Control', *Man-Environment Systems* Vol. **12** Issue **4** (1982), 121–160.
Hooker, C. A. (1983) 'The Future Must be a Fantasy', in Chapman, R. J. K. (ed.), *The Future*, Public Policy Monograph, University of Tasmania, 1983.
Hooker, C. A. (1986a) *A Realistic Theory of Science*, N.Y., State University of New York Press, 1986.
Hooker, C. A. (1986b) 'Evolutionary Naturalist Realism — Circa 1985', in Hooker 1986a.
Lewontin, R. C. (1978) 'Adaptation', *Scientific American*, Sept. 1978.
Lewontin, R. C. (1982) 'Organism and Environment' in Plotkin, H. C. (ed.), *Learning Development and Culture*, John Wiley & Sons, 151–170.
Munevar, G. (1981) *Radical Knowledge*, Indianapolis: Hackett.
Trigg, R. (1980) *Reality at Risk: A Defense of Realism in Philosophy and the Sciences*, Brighton, Sussex, Harvester Press.
van Fraassen, B. C. (1980) *The Scientific Image*, Oxford: Clarendon Press.
Waddington, C. H. (1975) *The Evolution of an Evolutionist*, Edinburgh: Edinburgh Univ. Press.

LARRY LAUDAN

ARE ALL THEORIES EQUALLY GOOD? A DIALOGUE

This essay is written as a dialogue between a relativist and his critic. It does not focus on all species of relativism (e.g., I do not directly address here either ontological relativism, cultural relativism or moral relativism), but specifically on what might be called epistemic or cognitive relativism. I understand that view to amount to the claim that we are never warrantedly in a position to assert that any theory is objectively superior to another, or that the evidence and arguments in favor of one theory are stronger than those favoring any other. Although the form of relativism I am describing is perhaps most closely associated with Thomas Kuhn[1] and Paul Feyerabend[2] among philosophers of science, closely-related versions of it can be found in the writings of Mary Hesse,[3] Willard Quine, Richard Rorty, Nelson Goodman,[4] Gerald Doppelt,[5] and sundry sociologists of knowledge.[6] The dialogue takes literary license in imaginatively elaborating some of the argumentative details of the relativist's position (in many cases, the relativists offer sloganistic hand-waving where argumentative fine-structure is called for). But I believe that all the major argument-types which I attribute to the relativist can be found in the writings of these authors.

1. INTRODUCTION

Relativist (hereafter: *R*): It is sometimes charged by our ubiquitous critics that relativism is an elusive account of science or knowledge because its central theses seem to be constantly shifting and undergoing re-definition. Being both clear- and hard-headed by nature, I want to allay *that* suspicion by setting out a version of relativism about science which is both clear in its assertions and cogent in its argumentative grounding. Although more elaboration will be required (and I stand prepared to give it if necessary), I shall begin by giving you a succinct formulation of the core thesis of epistemic relativism as follows: theories emerge in a certain cognitive or cultural milieu. Once a theory has been 'legitimated' within a particular context of ideas (viz., once it has come to be believed by a community of inquirers), there is nothing

Robert Nola (ed.), Relativism and Realism in Science, 117–139.

— neither evidence nor argument — which could compel a rational scientist to abandon that theory. It follows as a natural corollary that, if it comes to a choice between two or more rival theories which emerged in different contexts ('contexts' here can refer to units as diverse as paradigms, cultures, linguistic communities or 'forms of life'), then such a choice will inevitably be arbitrary. In sum: no rational scientist need ever give up any theory he is fond of.

When scientists do abandon their theories, and sometimes they do, it is not because logic or the evidence drives them to change their minds, but simply because they decide to abandon one theory and accept another. If you don't mind my putting it this way, the ecology of theoretical belief change is much closer to what you find among trendy Southern Californians than it is to the textbook picture of the scientist as rational decision-maker.

Non-Relativist (hereafter: *NR*): Thus far you have given us not arguments but assertions . . .

R: All in due course. As I shall try to show, the core arguments for the claim that a scientist is never forced to abandon his theories are quite simple. They boil down to these two: (1) as Duhem and Quine argued, theory choice is always radically underdetermined by the relevant data; and (2) as Kuhn showed, such shared standards of theory appraisal as there are always turn out to be highly ambiguous.[7]

NR: I'll turn in a minute to hear your arguments about underdetermination and ambiguity, but first I want to get a little clearer about the doctrine you are proposing to defend. Do I take it that you understand relativism to be the thesis that there are *never* objective grounds for choosing between *any* two theories?

R: Exactly, with the proviso that this applies only to theories which have been legitimated within some particular context or practice.

NR: That seems to be rather an extreme version of relativism. Wouldn't it be easier to defend the claim that there are *certain* rival theories between which rational choice is impossible?

R: Of course it would; but I don't adopt my beliefs with a view to making them easy to defend. The trouble with the thesis you just propounded is that virtually no serious scholar would deny it.[8] For instance, even the non-relativist would concede that we are not now in any position epistemically to choose between rival theories about what happened before the big bang. Neither are we now in any position to choose rationally between the theory that certain supernovae remnants came from explosions of Cepheid variables and the theory that they did not. If relativism is to have any teeth in it, it must assert that *all* genuine theories are on a par epistemically. More importantly, if the relativist commits himself to anything less than such a thesis, then he is, in effect, acknowledging that *some* theories are objectively superior to certain others. Such an admission would undermine the force of the relativists's challenge to the objectivist element of traditional epistemology.

Moreover, you should bear in mind that several prominent relativists defend precisely such a view. Thomas Kuhn, for instance, asserts that a scientist working in a particular paradigm can never be confronted with arguments and evidence which would force him to abandon the theories associated with that paradigm.[9] That claim is perfectly general, intended to apply to any rival paradigms or rival theories under consideration. (Cultural relativists such as Peter Winch apply a similar thesis to non-scientific theories as well.) If Kuhn were once to admit that there are some circumstances when a scientist might be rationally forced to give up the theories associated with his paradigm, then his wholesale argument for the immunity of inter-paradigmatic disputes to rational adjudication would collapse. So, as you can see, either one defends the thesis that all serious theories can be held onto come what may or else one abandons robust versions of relativism.[10]

NR: But in that case, I don't see that anything significant separates the relativist from the skeptic, and we both recognize that skepticism is self-defeating.

R: I prefer to leave the skeptic to speak for herself. But I do take exception to your assimilation of relativism to skepticism. Indeed, they are about as far removed as any two epistemological doctrines can be. The skeptic says that all theories are equally bad, because each can be disproven. The thorough-going relativist like me insists that all theories are equally good, because no theory can ever be disproven or decisively discredited.[11] If you have arguments to make against relativism, let us hear them; but do not resort to guilt by association.

2. THE ARGUMENT FROM UNDERDETERMINATION

NR: You said earlier that two of the central arguments for relativism derive from an understanding of the implications of theory underdetermination and from a realization of the ambiguity of all standards of evaluation. Let's deal with those claims in turn.

R: Precisely. The famous Duhem-Quine thesis shows that all theory choices are radically underdetermined by the evidence, no matter how extensive that evidence may be, and no matter what the theories in question happen to be. As Quine and more recently Mary Hesse have argued, relativism about theory choice follows as a natural corollary.

NR: Not so fast. You said that theory choice is radically underdetermined by the evidence. I don't mean to nit pick, but can you tell me what 'radical' underdetermination is, as opposed to the garden-variety sort? Do you mean that it is worse, or more extreme than ordinary underdetermination, or that it is more threatening to traditional epistemology? I'm really just asking for a clarification here.

R: Suppose that we simply settle for 'underdetermination' and leave Quine and others who prefer the stronger language to explain what they mean. Even without the emphatic adjective, underdetermination shows the hopelessness of methodology.

NR: I want to return to your last remark but one. You claimed that Duhem and Quine showed relativism (by which you mean that an established theory can never encounter evidence which requires rational scientists to abandon it) follows from underdetermination. But it has always seemed to me that what Duhem and Quine actually showed was that the falsity of a theory could not be deduced from the failure of that theory to make a correct prediction, since collateral or auxiliary assumptions were always involved in deriving the prediction. As those collateral assumptions might be responsible for the error, it follows that the falsity of the theory cannot be deduced from the falsity of one of its predictions.

R: Exactly, which is just to say that no body of evidence can force the rejection of any theory.

NR: It is to say rather that no set of evidential statements entails the falsity of a theory.

R: Well, they come to more or less the same thing.

NR: They don't if you'll grant that there's a possible difference between the logic of entailment and the causal generation of propositional attitudes. The fact that I believe a theory false may or may not be grounds to reject it. If scientists rejected all the theories which they believed to be false, little theoretical science could be taught (especially if you consider the role of idealizations, viz., known falsehoods, in physical theory[12]). Hence the determination that a theory is false and the decision to reject it are two quite different matters.

R: Well, what I meant to say was that a scientist, knowing that a theory was one part of an inference which led to a false prediction, need never feel forced to reject the theory, just because it was 'implicated' in a failed prediction. Mary Hesse showed as much in her recent *Revolutions and Reconstructions in the Philosophy of Science.*

NR: But Hesse's argument, and yours, rests on the monumentally dubious assumption that the rejection of a theory is warranted *only if* the theory has been logically falsified, i.e., if the only relevant rule for theory rejection is some version of *modus tollens.* Indeed, the whole *brouhaha* about underdetermination seems to rest on the idea that the only rule of rejection open to the scientists is *modus tollens.* But at least since Hume we've known that the laws of deductive logic were insufficient for doing empirical science (at least all of us except Popper realized that). Customarily, the laws of deductive logic are supplemented by a wide variety of ampliative criteria of theoretical acceptability. We commonly insist, for instance, that acceptable theories must be general in scope, simple, and that they make surprising predictions. *The arguments of Duhem and Quine — at most — establish only that theory choice is underdetermined by the rules of deductive logic*, not that they are similarly underdetermined by the rules of a richer ampliative logic of theoretical inference.

R: That may be so, but surely you are ignoring the fact that Nelson Goodman showed that the rules of enumerative induction, just like those of deduction, underdetermine theory choice.[13] For instance, he pointed out that no rules of enumerative induction can allow us to choose between the claim that "All emeralds are green" and the claim that "All emeralds are grue."

NR: Well, although you've got the example right, that's not quite what Goodman showed. He argued specifically that there are certain contrary hypotheses between which the law of enumerative induction could warrant no choice. But that is a far cry from asserting — as you do — that inductive methods never justify a choice between any two contrary theories. For instance, if I propound the hypothesis that "All emeralds are blue", and if I then observe an emerald which is green now, I can, using the rules of inductive logic, reject that hypothesis. Nowhere does Goodman show that the rules of enumerative induction underdetermine choice between *any* two rival hypotheses, only that they do so between a very delimited set of rivals.

But we can leave the finer points of Goodman exegesis to another time, because the larger mistake you make is that of assuming that Goodman's argument — even if it showed that enumerative induction, like *modus tollens*, always underdetermined theory preferences — would establish your thesis that *all* theories are evidentially on a par. Science clearly involves a much wider set of criteria of theory appraisal than the straight rule of induction, which alone has attracted Goodman's attention. Since it does, the alleged inability of the straight rule to enable us to make choices between theories leaves completely open whether the larger panoply of methods in the scientist's repertoire similarly has this feature. Unless you can direct me to the relevant literature, I incline to the belief that *no one has ever shown, although many have alleged, that the ensemble of methods we normally call 'scientific' underdetermines theory choice.*[14]

R: But many of your fellow critics of relativism (such as Boyd[15] or Newton-Smith[16]) concede that underdetermination poses a serious challenge to their position.

NR: And so it does, for the scholars you mention are epistemic *realists* who want to be able to show that their theories are true or close to the truth. Such limited logical underdetermination as Quine and Duhem identified does pose at least a *prima facie* problem for realists, who think that we are justified in accepting (or rejecting) a theory only when we can reasonably assume that it is true (or false). But since I am no more a realist than a relativist, I can happily grant that underdetermination may be the undoing of realism without granting that it provides the slightest aid and comfort to relativists.

R: Now at least one of us is confused. If you accept the point that we are never in a position warrantedly to assert the truth of a scientific theory, or even its approximate truth, then you are perforce a relativist. What indeed are we quarreling about?

NR: You make a common mistake, which is no more excusable for its ubiquity. The doctrines of realism and relativism may be mutually exclusive, but they are not jointly exhaustive. So both might be wrong. Because that is so, weaknesses in realism, and there are plenty, provide no argument for relativism and vice versa.

R: So you think that there is a coherent middle ground between the two?

NR: It's not really 'middle ground', if by that phrase you have in mind an eclectic combination of elements of each. What I am saying, rather, is that the debate about realism is orthogonal to the debate about relativism. The *realist* and the *non-realist* debate about whether one is generally entitled to believe that one theory is 'truer' than another. Here, I suspect that the non-realist has the stronger argument. But what the *relativist* and the *non-relativist* dispute is whether one is ever entitled to hold that one theory is objectively 'better' than another (where 'better' need not mean 'truer' but something like 'better tested' or 'better supported').

To see why you need to do more than defeat the realist, consider your own position as you described it earlier. As a relativist, you hold that one can never be rationally compelled to abandon a theory in favor of a rival, since there are (you say) no objective grounds for holding one to be superior to the other. If your point is to have any punch to it, 'superior' here must have a broader sense than 'truer', since you presumably think that no cognitive appraisals of a theory (e.g., that it is more reliable than its rivals, that it is more intelligible, that it is less *ad hoc*, etc.) are rationally adjudicable. By contrast, I believe that there are often cases when it is rational to hold that one theory is 'better' than another (relative to some cognitive utility structure), even though both theories may be, so far as we know, false. And I now reiterate my claim that no one has shown that the methods actually used by scientists underdetermine theory choice in the sense you require.

3. THE ARGUMENT FROM THE AMBIGUITY OF STANDARDS

R: But many of these methods and rules you are now speaking of are themselves so ambiguous and amorphous that they could be used to justify more or less any theory choice whatever. That, after all, is what Kuhn showed in chapter 13 of *The Essential Tension.* Consider, for instance, his point that simplicity (or *adhocness*) has such a multiplicity of meanings that virtually any theory could be said to be 'simple' or '*non-ad hoc*' in some sense or other. Feyerabend too, I seem to remember, showed that methodological rules have no teeth in them. Specifically, Feyerabend showed that scientists are constantly violating your hallowed methodological rules.

NR: Hold on. You can't have it both ways. If you maintain with Kuhn that standards are so ambiguous as to favor any theory, you can hardly also claim (with Feyerabend) that those standards are honored more in the breach than the observance. If the rules are really as ambiguously permissive as you and Kuhn claim, then how can you also hold that scientists routinely violate them?

R: It's not so simple as that, but since it is the ambiguity of rules which is now under discussion, let's stick to that point.

NR: And so we should. It's true that Kuhn made much hay with the notorious ambiguities of such criteria as simplicity and *adhocness*. But consider other criteria, including those that Kuhn himself mentions. Scientists, he says, expect theories to be consistent, to be highly general in scope, and to make surprising predictions.[17] Kuhn never shows that any of *these* evaluative criteria are so ambiguous that virtually every theory could be shown to satisfy them. Quite the contrary, we can point to plenty of scientific theories which were rejected because they were eventually seen to be inconsistent, or insufficiently general or lacking in novel predictions.

R: But scientists don't always reject theories just because they appear inconsistent or because they've failed to make novel predictions.

NR: That's surely true. But that does not establish the *ambiguity* of the notions of consistency or surprising predictions. It shows rather that scientists are sometimes willing to waive the normal standing rules they profess. Whether Feyerabend is right that it is always appropriate to waive such standards is yet a different question, perhaps for another occasion. For now, we are dealing with your (and Kuhn's) claim that the rules of theory choice are too vague ever to warrant a choice between rival theories. I've given you examples of methodological rules which are decidedly unambiguous.

R: But even when they are unambiguous, these methodological rules may be unavailing. Suppose we are confronted with two rival theories, each of which is internally consistent or each of which makes surprising predictions. In such circumstances, your much-vaunted rules of choice are powerless.

NR: There will, of course, be circumstances where one or another of these methodological constraints will be satisfied by more than one extant theory. (If this never happened, then we might expect scientists never to disagree with one another.) But that is a far cry from saying — as your position commits you to saying — that in all circumstances, every extant theory will satisfy all the relevant constraints. Here, as elsewhere, you seem to have difficulty distinguishing existential quantifiers ("for some theory rivals", "rules are sometimes unavailing") from universal quantifiers ("for all theory rivals", "rules are always unavailing").

4. THE ARGUMENT FROM THE CONVENTIONALITY OF STANDARDS

R: Even if I were to grant you that methodological rules are sometimes specific enough to determine a theoretical preference, it will not strengthen your case against relativism, for the simple reason that the methodological rules, which you set such store by, are themselves just conventions. In other words, even if a certain rule or set of rules, M, indicates that T_1 should be preferred to T_2, that makes the choice of T_1 no less arbitrary, since the use of M is without rationale. If you have doubts on this score, look at the writings or your fellow anti-relativists like Popper or Lakatos.[18] They go to some pains to stress that methodological rules are simply conventions for playing the game of science. As conventions, methodological rules can have no epistemic or cognitive bite. They do not rest on any determinable facts of the matter. Hence their invocation to settle scientific disputes can only beg, rather than settle, questions of epistemic warrant.

NR: You are right that the main-line tradition in 20th century philosophy of science has tended to regard methods as matters of convention. For good measure, you might have added Carnap and Reichenbach to the list of advocates of that bizarre position. You are also right that strict conventionalism about methods is a well-greased slope leading directly to strong relativism. But even the slightest reflection will show that methods of science need not be regarded as arbitrary conventions, and that the positivists were flatly wrong to regard them as such.

If methodological conventionalism were right, then we should expect that a scientist (or philosopher of science), when challenged about the soundness of a proposed methodological rule, would reply by shrugging his shoulders and saying that it's just a matter of taste. But that is not the way it goes. Rather, one justifies the selection of that method or rule by showing that it is an optimal means for realizing our basic cognitive ends or utilities. Just as our rules pick out theories, so do our aims or ends pick out the methods we should use. In my view, the claim that a method is sound is equivalent to the claim that it promotes our cognitive ends; and that in turn is just a factual claim about the efficacy of realizing those ends. If you'll grant me the plausibility of claiming that flapping one's arms is not an effective way to fly, then I'll argue for the plausibility of claims of the form: "Using M is (or is not) an effective way of bringing about A". Methodological rules are just means to our cognitive ends and they can be assessed in the same way that any other claim about ends and means can be.

R: So you are saying that methodological rules are testable in the same way that theories are?

NR: More or less, from which it follows that methodological rules are *no more conventional than scientific theories.* That remark will not cut much ice with you, since you think that theories are just conventions as well. But I do want to insist that theories and methodological rules are on a par in terms of their empirical and contingent character. Anyone who holds that scientific theories are non-conventional and that methods are conventions is speaking incoherently.

5. THE ARGUMENT FROM AXIOLOGICAL RELATIVITY

R: You can play this temporizing game if you like, always invoking higher levels of justification, whenever I show the arbitrariness of decisions at one level. But surely you can see that, on pain of infinite regress, you must stop somewhere. If the aims of science come at the top of your justificatory hierarchy, as it seems they do, then all my arguments about the arbitrariness and conventionality of methodological rules apply with a vengeance to aims. I hold that there can never be non-arbitrary grounds for claiming that *any* proposed aim for inquiry is any better or more appropriate than any other. And the whole of the positivist and empiricist tradition — from Hume to Hempel, Carnap, Reichenbach and Popper — sees things exactly as I do.[19] Now, since your choices of method and thus ultimately all your theory choices evidently depend on your cognitive aims or standards, the whole system is mired in arbitrary and subjective whims. You can make the circle larger, but you can't extricate yourself from it.

NR: Your remark about the relativism implicit in positivism and logical empiricism is quite well taken. Despite their avowed opposition to relativism, all of the leading figures of this movement turned out to be unrepentant subjectivists about axiological questions. Because they were, and because all scientific choices amount to decisions which inevitably involve the use of cognitive values of one sort or another, they end up more nearly on your side than on mine. But before I can show you why both you and the positivists were too hasty about these matters, I'd like to bring our discussion to ground by making it a bit more specific and concrete. Let's go back to your initial claim which was, if I recall, that there could never be good reasons for a scientist to give up one theory and accept a rival. Am I right in thinking that you still hold to that view and that the strongest argument you can adduce for that claim is that all theory choices ultimately depend on invoking cognitive aims which are themselves not open to rational negotiation?

That, in short, axiological relativism is what ultimately undergirds your epistemic relativism?

R: In a nutshell, yes. Nor am I alone in seeing relativism rooted in the subjectivity of cognitive values. After all, Kuhn's central argument for the relativism of paradigm choice is really the claim that the advocates of different paradigms subscribe to different cognitive standards and — like the positivists — Kuhn believes that cognitive standards are not negotiable, not criticizable.[20]

NR: I want to persuade you of two things. First, that even if cognitive values are not 'negotiable' (as you put it in your characteristically cynical way), it does not follow that scientists could never have very good reasons for giving up one theory and embracing another. But secondly, and more importantly, I want to show you that there is scope for rational deliberation about basic cognitive ends.

R: I doubt that you can persuade me of either.

NR: Let's begin with the easier case first. Suppose a scientist who subscribes to a methodology, M_1, picks theory T_1 because, according to M_1, T_1 is better than T_2. Suppose further that new evidence emerges in the course of that scientist's work. As he reassesses his theory preferences in the light of that evidence, are there no circumstances under which he might be forced to the conclusion that — according to the M_1 — he must give up T_1 and accept T_2?

R: But you're assuming that his methods fail to underdetermine his choice between T_1 and T_2.

NR: So I am, but we've already been over that ground. I alleged, and you did not deny, that no one had shown that methodological constraints actually used by scientists underdetermine all theory choices. So, unless you're prepared to give me such a proof now, you should grant me that things might be as I have described them.

R: Under mild protest, proceed.

NR: So, a scientist might conceivably have good reasons for giving up T_1 and accepting T_2; specifically, when he subscribes to a method, M_1, which optimally promotes his cognitive aims and when M_1 makes clear that T_2 is better than T_1. Isn't that a good reason for preferring T_2?

R: Whether it's a *good* reason has yet to be established, although I grant you that it's a reason of some sort. But my original claim was really directed at the following concrete and commonplace situation: suppose we have two scientists, S_1 and S_2, who have *different* aims, A_1 and A_2, and thus different methods . . .

NR: Let me interrupt for a moment. I think I just heard a familiar connective in a strange place. You asked us to consider a situation where different scientists had different aims and you went on to say that they would 'thus' have different methods. Why does it follow as a matter of course that scientists with different aims will automatically subscribe to different methods?

R: Well, I haven't thought all that much about it, but isn't it true that philosophers as different as Popper and Kuhn assume as a matter of course that different aims will be promoted by different methods? Surely it's pretty obvious.

NR: Obviously false, that is. Scientists with very different aims can sometimes agree about which methodological rules to follow because it may happen that one set of methods turns out to be instrumental in furthering quite diverse aims. That is the reason why realists and instrumentalists (e.g., Mach and Boltzmann) in science are so often able to agree about which methods of theory appraisal to use, even when they disagree deeply about the axiology of science.

R: Well, never mind, because the 'thus' was not crucial to my argument. Let me again try to state my point. Suppose we have two scientists, S_1 and S_2, with aims A_1 and A_2 respectively, and suppose that M_1 promotes A_1 and M_2 promotes A_2. Suppose, finally, that M_2 picks out theory T_1 while M_2 picks out a rival T_2. In short, we have two scientists who subscribe to different 'paradigms'. Under those circumstances, I hold that nothing could ever rationally persuade S_1 to abandon T_1 and accept T_2.

NR: You're moving a bit quickly here. Isn't it possible that even if S_1 holds firmly to his aims, A_1, that he might reasonably come to believe that some methodology M^* rather than M_1 optimally promotes A_1? After all, people do change their beliefs about ends/means connections in the face of new evidence and arguments.

R: Perhaps . . .

NR: And isn't it conceivable that M^* may pick out T_2 as better than T_1? In which case, S_1 would have good reasons for picking T_2 over T_1. Indeed, he would be inconsistent to do anything else. So, on pain of inconsistency, he would have to change his mind about which theory to accept.

R: I suppose it's conceivable. But you perversely continue to evade the upshot of my argument. The cases that really interest me are those where S_1 believes that A_1 is optimally promoted by M_1 and where M_1 picks T_1. Here, I claim, there could never be overwhelming rational grounds for S_1 to abandon T_1 and accept T_2.

NR: Let's recapitulate. You began this discussion by claiming that there could never be arguments which would force a rational scientist to give up one theory and to accept another. I first pointed out that new evidence might lead S_1 to accept T_2, because S_1's methodology requires this.

R: Yes, but so long as S_1's methodology picks T_1, nothing could budge him.

NR: But we saw that there were circumstances where M_1 itself might be rationally abandoned to serve S_1's cognitive aims; and that abandonment in its turn might lead to S_1's accepting an M^* which mandated the selection of T_2.

R: Such events might occasionally arise; but as long as S_1 believes that M_1 promotes his A_1 and so long as M_1 picks T_1, then S_1 could never be persuaded to accept T_2.

NR: And you believe that this state of affairs is common in science?

R: Not merely common, but the general rule when – as in cases of inter-paradigmatic confrontation — scientists espouse different aims or goals. Indeed, the major trouble with you rationalists is that you have failed to come to terms with one of the central lessons from the history of science and philosophy; namely, that scientists in different epochs have different cognitive agendas. Still worse, scientists of the same epoch who work in different research programs also typically subscribe to different cognitive utilities.

So long as epistemologists could assume that all scientists were after the same thing (typically cashed out as "true knowledge for its own sake"), then they could persuade themselves that — although scientific theories themselves changed, and perhaps even the methods of science — the basic aims of the enterprise had remained unchanged since man started gathering nuts and berries. Indeed, the positivists could avoid having to face up to the consequences of their relativism about aims by imagining — as some evolutionary epistemologists do nowadays — that all scientists have precisely the same aims. What we relativists have managed to do is to come to terms with the implications of the existence of rival aims and divergent standards among scientists.

NR: I grant you that non-relativists have been slow to admit the reality of shifting epistemic standards, and many (e.g., Lakatos and Worrall) are still reluctant to admit it. Philosophers are especially culpable here, since the history of their own discipline should have made it clear to them that it was implausible to think that the aims of inquiry were either univocal or unchanging.[21] But I fail to see why, even if we grant your point about divergences in standards, it follows that the relativist is home free.

R: Let me take your query as an opportunity to restate my position. My claim about cognitive egalitarianism boils down to this: in cases where

(1) S_1 and S_2 have different cognitive aims, A_1 and A_2 respectively; *and*
(2) S_1 and S_2 espouse different methods, M_1 and M_2 respectively; *and*
(3) it is not possible to persuade S_1 that M_1 is suboptimal for realizing A_1; *and*
(4) where M_1 picks out T_1 and M_2 picks out T_1; *then*

(5) there is no hope for rational closure of the disagreement about T_1 and T_2.

NR: It's a relief that you have finally formulated relativism in a form which is reasonably perspicacious. Would it be fair to assume that you believe that conditions (1) through (4) are fairly typical?

R: Quite. And do you deny that such cases ever arise?

NR: No actually, I don't. I do not think that they are as ubiquitous as you evidently do; but I concede that this state of affairs sometimes arises.

R: At least you've learned something from history! Since we both agree about these facts of the matter, and since it is the existence of such cases which provides the core rationale for my position, I think it's important for us to understand why an impasse of this sort is reached by scientists. In the circumstances I have described, the only way to talk S_1 or S_1 out of their theory preferences would be by showing that their aims are wrong or inappropriate. And it is precisely the possibility of discrediting such cognitive aims that I am denying. Moreover, this situation is especially apt to arise in science in connection with disagreements about theories between the advocates of rival paradigms; this is why Kuhn rightly observed that debate between proponents of rival paradigms is always inconclusive.

NR: So, it is the lack of a decision procedure for axiologies, rather than your earlier arguments about underdetermination and ambiguity, where you now wish to root your epistemic relativism.

R: Precisely. And, as I said to you earlier, I am here making common cause with Popper, Reichenbach, Carnap, Duhem and most of the other folk heros of 20th-century philosophy of science. I make only three assumptions: (a) that axiological claims are non-negotiable matters of taste; (b) that, where axiological differences do exist, they are apt to lead to different sets of methodological and theoretical preferences; and (c) that different scientists in fact subscribe to different axiologies. Assumptions (a) and (b) are already widely assumed by theorists of rationality and the historical record provides ample evidence for (c). Taken together, (a), (b) and (c) entail epistemic relativism.

NR: So they do. But before we pronounce relativism victorious, let's examine one of the core premises, (a), in your derivation. Specifically, let's talk a bit about what having an aim amounts to. I suppose, to begin with, that we are talking specifically about cognitive aims, i.e., about the aims of science; and more specifically, we're talking about the aims behind theorizing. Would it be reasonable to say that a cognitive aim identifies certain properties or characteristics that we want our theories to have?

R: Obviously. And scientists in fact differ about what those should be. Some scientists have the aim of finding true theories about the world; others are seeking to save the appearances; other are looking for theories which will solve particular sorts of problems . . .

NR: But linger a while over this issue before we confront the question of differences between scientists. A cognitive aim thus states that a certain attribute, or set of attributes, is desireable in our theories. And to have, or to adopt, such an aim is just to seek to create and accept theories which exhibit such attributes.

R: Of course. An aim would be no aim at all unless it served to guide the scientist in his selection of theories.

NR: So on your view, it would be a contradiction to say that a scientist had a certain aim but was unconcerned to engage in actions which would promote or realize that aim?

R: You could put it that way.

NR: So if a scientist were to come to believe that a certain aim was unrealizable, then he could no longer be said to have that aim?

R: I suppose so; but all this is beside the point since it is universally assumed that one could never show that an aim is unrealizable.

NR: What if a scientist has a certain aim which he comes to believe is inconsistent?

R: How can aims be inconsistent? Aims are not statements; they are value judgments. The logic of consistency and inconsistency applies only to statements. Since aims are neither true nor false, they cannot be said to be inconsistent.

NR: Let me put the point more carefully. Suppose that a scientist has aims A_1 and A_2, and he is subsequently persuaded that A_1 can be realized only in circumstances where A_2 fails to be realized, and vice versa. In other words, the scientist comes to believe that theories could possibly realize one or the other of the aims he specifies, but not both. Call that state of affairs 'axiological inconsistency'. Now, wouldn't the discovery that one's aims were 'inconsistent' in this sense force a rational scientist to abandon at least one of his aims?

R: Only if he had a more basic aim of being consistent. But then that aim would be as arbitrary as the rest.

NR: You talk about the desire for consistency itself being an aim, and thus (on your view) a matter of taste. But I thought you had earlier agreed that it was *constitutive* of having an aim or goal that one was prepared to engage in actions designed to promote that goal. In that case, acknowledging that one had an inconsistent axiology would be tantamount to admitting that one had goals which there was nothing in principle one could do to bring about. And isn't that tantamount to recognizing the necessity for changing one's goals?

R: It does seem that inconsistent goals or aims could never serve as the basis for any set of actions and thus it would be quixotic for an agent to retain what he acknowledges to be 'inconsistent' goals.

NR: Fine, now we're getting somewhere. So a set of cognitive aims can be legitimately faulted if it can be shown to be inconsistent.

R: Yes; but it is surely very rare for one to be able to demonstrate that a set of aims is inconsistent.

NR: Whether it is rare is an empirical matter, which I should like to leave for later. My concern here is simply to establish the in-principle possibility that scientists might be rationally compelled to change their aims. Nor is inconsistency the only factor which can act as the engine of axiological change. Consider, for instance, the fact that a scientist might come to believe that his aims, although perfectly consistent with one another, are not realizable.

R: How could such a result ever be established?

NR: Consider a well-known example from the early 19th century. For a very long time, one of the central aims of science was the establishment of theories with apodictic certainty. Ever since Aristotle's *Posterior Analytics*, prominent scientists and philosophers had adhered to the ideal which said that scientific knowledge was fully certain and infallible. By the early 19th century, thanks in part to the arguments of Hume but also because of the overthrow of some widely-believed theories (e.g., Newtonian optics), most scientists and philosophers came to be fallibilists, and to hold that certainty could not possibly attach to the generalized conclusions of empirical research.[22] What we have here is a case where scientists came to believe that a cognitive aim which had long been thought to be central to science was not realizable. As a result of that discovery, reflective scientists and philosophers ever since have refused to construe the aim of science as the development of absolutely certain theories. I'm curious what you make of this case, since I gather that you yourself are as much a fallibilist as the rest of us.

R: I am a fallibilist, but not for the reasons you evidently think. My personal tastes favor seeking theories which are useful and which save the phenomena; but, on pain of contradiction, I cannot hold that my tastes are better than anyone else's.

NR: Leave that issue aside. What I am asking you is whether you agree with me that the abandonment of infallibilism and of 'the quest for certainty' was the only appropriate response to the critique of infallibilism.

R: It may have been a response to that critique, but it was neither reasonable nor unreasonable. Scientists simply decided that they wanted to change their epistemic aims, but there could have been nothing which rationally compelled them to.

NR: But you've already conceded that an unrealizable aim is no aim at all. Since we're agreed that having an aim which is realizable is more or less constitutive of goal-directed action, don't you see that unrealizable aims are irrational?

R: I see no such thing. Many agents, perfectly rational ones at that, have had a variety of aims which turn out, so far as we know, to be totally unrealizable. Alchemists wanted to turn lead into gold; astrologers sought to predict human fortunes from the positions of the planets . . .

NR: Your correction is quite well taken. Rational agents may have unrealizable goals, so long as they have not seen the evidence and arguments for their unrealizability. But once they see that their aims are unrealizable, it no longer becomes coherent for them to hold onto them as aims. Return to my example of the abandonment of infallibilism. You're right that there was nothing irrational about Aristotle, Newton or Locke holding that science aimed at certainty in knowledge. But once Hume *et al.*, had managed to convince scientists that there was no way of producing apodictically certain theories, doesn't it follow that those scientists should give up certainty as an aim of inquiry?

R: It seems to me that something very fishy is going on here.

NR: On the contrary, the argument is quite simple and quite traditional. All I am asking you to concede is the point — as philosophers were once wont to put it — that 'ought' implies 'can'. To say that scientists ought to be aspiring to realize a certain aim presupposes that scientists are in a position to be able to bring that aim about. If a scientist comes to believe that a certain aim is inherently unrealizable, then he can neither urge others to adopt it nor rationally retain it as a guide to his own actions. We have already agreed that aims are meant to serve as guides to action. Once we come to believe that no actions whatever will realize or promote a particular aim, then it can no longer function as an aim.

R: I grant you that it seems peculiar to imagine a scientist saying "My aim is x, and I firmly believe that x cannot be realized." But even if I concede the point in principle that occasionally there can be such grounds for criticizing an aim, you surely are not expecting to get very far with this concession. Most of the aims about which scientists disagree are perfectly realizable; and so this bit of critical machinery is of little or no use.

NR: Quite the contrary, very many of the aims seriously entertained by past and present scientists can be criticized on precisely these grounds. I already showed you that it was in just this manner that scientists gave up the quest for certainty in the 19th century. It is on very similar grounds that Popper showed the unrealizability of the aim of finding theories which have a high degree of probability. Others have recently criticized the whole cluster of aims associated with scientific realism precisely on the grounds that those aims are unachievable[23] . . .

6. CONCLUSION

On the verge of becoming rancorous and repetitive, the dialogue breaks off here. What has been established thus far is that there is a wide variety of circumstances in which rational consensus about theories could at least in principle be established. This is manifestly not to say that all theoretical disputes can be resolved, nor that all axiological differences can be brought to rational closure. But granting the non-resolveability of *some* theoretical and axiological disputes should provide precious little comfort to the relativist, since his wholesale cognitive egalitarianism about theories commits him to the non-resolveability of *all* such disputes.

And once we realize that any particular dispute about theories or standards may be resolveable, we can see that the relativist's knee-jerk reaction to every controversy, namely to issue a hearty *de gustibus non disputandum est*, is phony through and through. The interesting question is *which* theoretical and axiological disputes are resolveable and which are not. My deepest disagreement with the relativist centers on his conviction that this issue can be settled *a priori* and wholesale by philosophical legislation or fiat. In my view, it must always be settled in terms of a detailed knowledge of the case in hand, and each case must be separately argued for. If classical rationalism over-reached itself in holding that *all* disputes about matters of fact could be rationally resolved, contemporary relativism — with its insistence that *none* of them can be — is just as pernicious and, at least in our time, far more widespread.

University of Hawaii

NOTES

[1] See both Kuhn (1970) and Kuhn (1977).
[2] See especially Feyerabend (1975).
[3] M. Hesse (1981).
[4] See especially Goodman (1978).
[5] G. Doppelt (1978).
[6] Especially Barry Barnes, David Bloor and Harry Collins.
[7] See especially Kuhn (1977), ch. 13.
[8] G. Doppelt (1986) goes to great pains to show that there are *some* rival theories such that the relevant evidence and rules of appraisal do not now permit a rational choice between them. (He calls this thesis 'weak relativism'.) If this were not so, then there could never be rational disagreements between scientists! Doppelt's 'weak relativism' is a truism about science, and it provides no warrant for the strong relativist claim (which Doppelt himself once espoused) under discussion here. For a detailed discussion of these matters of strong and weak relativism, see my 'Relativism, Naturalism and Reticulation', *Synthese* **71** (1987), 221–34.
[9] There are many threads of argument which lead Kuhn to this conclusion. Chief among them are his claims that:

(1) the change from an old to a new paradigm involves a *change of standards*, but not a raising of standards [Kuhn (1970), p. 108].
(2) the arguments offered for every paradigm are circular (*ibid.*, p. 94).
(3) inter-paradigmatic communication is 'ineffective' (*ibid.*, pp. 147–9).
(4) scientific standards are ambiguous [Kuhn (1977), p. 321–2].
(5) acceptance of a paradigm is influenced by non-scientific factors (*ibid.*, p. 325).

[10] For many of the same reasons, Feyerabend's critique of scientific methodology would come unstuck if he defended anything weaker than the version of relativism under discussion here.
[11] You might say that the relativist is to Popper as the Humean sceptic is to the inductivist.
[12] For the relevant arguments, see Duhem (1954) or Cartwright (1983).
[13] See the classic monograph Goodman (1965).
[14] In saying this, I do not mean to imply that there is such a thing as *the* scientific method. However, I do believe that there are methods utilized by scientists (although not exclusively by them). For more details, see Laudan (1983).
[15] See Boyd (1973).
[16] See W. Newton-Smith (1978).
[17] Kuhn (1977), ch. 13.

[18] For Popper on the conventionality of methodological rules, see section 11 of Popper (1959); for Lakatos on the same theme see Lakatos (1978), p. 20 ff.
[19] See, for instance, Reichenbach (1938), pp. 10–13
[20] Cf. especially the formulation of Kuhn's position in Doppelt (1978).
[21] How one can look at the history of epistemology, which is a long series of deep disagreements about the aims of knowledge, and still imagine that all thinking people have the same epistemic aims is wholly beyond me.
[22] For a detailed discussion of this case see Laudan (1981).
[23] See, for instance, Laudan (1984), pp. 103–37.

REFERENCES

Boyd, R. (1973) 'Realism, Underdetermination and a Causal Theory of Reference', *Nous* **7**, 1–12.
Cartwright, N. (1983) *How the Laws of Physics Lie*, Oxford, Clarendon Press.
Doppelt, G. (1978) 'Kuhn's Epistemological Relativism: An Interpretation and Defense', *Inquiry* **21**, 33–86.
Doppelt, G. (1986) 'Relativism and the Reticulation Model of Scientific Rationality', *Synthese* **69**, 225–52.
Duhem, P. (1954) *The Aim and Structure of Physical Theory*, Princeton, Princeton University Press.
Feyerabend, P. (1975) *Against Method*, London, NLB.
Goodman, N. (1965) *Fact, Fiction and Forecast*, Indianapolis, Bobbs-Merrill.
Goodman, N. (1978) *Ways of Worldmaking*, Indianapolis, Hackett.
Hesse, M. (1981) *Revolutions and Reconstructions in the Philosophy of Science*, Brighton, Harvester.
Kuhn, T. (1970) *The Structure of Scientific Revolutions* (Second Edition), Chicago, University of Chicago Press.
Kuhn, T. (1977) *The Essential Tension*, Chicago, University of Chicago Press.
Lakatos, I. (1978) *The Methodology of Scientific Research Programmes: Philosophical Papers Volume 1*, edited by J. Worrall and G. Currie, Cambridge, Cambridge University Press.
Laudan, L. (1981) *Science and Hypothesis*, Dordrecht, D. Reidel.
Laudan, L. (1983) 'The Demise of the Demarcation Problem', in L. Laudan and R. Cohen (eds.), *Physics, Philosophy and Psychoanalysis*, Dordrecht, D. Reidel, 111–28.
Laudan, L. (1984) *Science and Values*, Berkeley, University of California Press.
Newton-Smith, W. (1978). 'The Underdetermination of Theories by Data', *Aristotelian Society* Supplementary Volume **LII**, 71–91.
Popper, K. (1959) *The Logic of Scientific Discovery*, London, Hutchinson.
Reichenbach, H. (1938) *Experience and Prediction*, Chicago, The University of Chicago Press.

FREDERICK KROON

REALISM AND DESCRIPTIVISM

1. INTRODUCTION

It is often thought that there is a tight relationship between the metaphysical issue of realism, including scientific realism, and the semantic issue of how terms refer. Where conceded, this relationship is then commonly described in the following terms: if you are a descriptivist about the mechanism of term-reference, you are almost bound to be a kind of antirealist about what terms stand for; if, on the other hand, you are not a descriptivist but a proponent of the theory of direct reference, you make room for a corresponding realism, or (stronger) you thereby show yourself to be a realist about the objects and kinds directly referred to. I shall call arguments for the first half of this claim (that descriptivism leads to antirealism) *descriptivism-to-antirealism* arguments, or *DA* arguments for short. These arguments can take a variety of forms, but among the most familiar are a set of three which I shall call the (a) existential, (b) the relativist, and (c) the 'veil of descriptions' *DA* arguments. The existential argument is the least radical, and in many ways the easiest to understand. It alleges that if one is a descriptivist about the referential mechanism of terms, standards for successful reference have been set too high. According to descriptivism, a term t has a referent only if (a weighted most of) the correlated descriptions or properties pick out just one object or non-empty set of objects (should it be a general term), so that if a large enough number of the descriptions are not true of anything the term fails to have a referent, or fails to apply to anything. Because the most central and vivid beliefs associated with a term frequently fit nothing, or too many things, there is a good chance that many existential claims will now turn out false on descriptivism. Hence descriptivists ought to be what I shall call, following Plantinga,[1] *existential* antirealists about the alleged referents of many terms from the designated class. The exceptions are given by terms embedded in mostly true theories or belief-contexts.

Thus consider past scientific theories, with their talk of celestial spheres, intelligible species, phlogiston, ether and the like. Descriptivists

Robert Nola (ed.), Relativism and Realism in Science, 141–167.

will predict — correctly enough, I have no doubt — that we ought to be existential antirealists about such entities, and all because the theories embedding the terms are substantially false. More radically, however, descriptivists will predict antirealism even in the case of such familiar scientific terms as 'mass' and 'momentum' as used in pre-Einstein physics, and all because older pre-Einstein theories about these physical magnitudes are substantially false. More radically still, our own theoretical ontology is also under threat. By Putnam's well-rehearsed meta-induction,[2] from the point of view of future 'mature' science our own current theories will appear substantially false, and hence (applying descriptivism) current scientific terms will appear to lack a referent. Descriptivism threatens in this way to lose not only the world of past scientists, but much of our own world as well.

As I have understood the doctrine, existential antirealism concerning some range of posits is compatible with the view that there is a mind-independent world that contains whatever our eventual true theory of the world says there is. So existential antirealism is compatible with a moderate form of realism. Existential antirealism thus understood should be contrasted with another much more radical variety of antirealism, the relativistic form found in Kuhn's and Feyerabend's[3] writings on theory-change and the theory-ladenness of observation. Here, too, descriptivism about reference seems to constitute the motivation, a point made by a number of authors concerned to show the coherence of realism in the face of the Kuhn-Feyerabend argument. I shall call this way of making the 'descriptivism-antirealism' connection the *relativist argument*. Robert Nola gives a clear formulation of the relativist argument in Nola (1980), which describes relativist antirealism as a doctrine similar to existential realism in that 'none of the terms of our theories ever manage to refer to what is really in the world, but different in that each paradigm is fitted out with its own world of entities relative to which the terms of the theory in the paradigm refer' (p. 328). He then goes on to claim that such a view depends on supposing that the terms in a paradigm are defined by implicit definition from the core of theoretical sentences embedding the terms, and hence depends on a thick form of the descriptivist doctrine that the referent of a term is taken to be whatever fits the relevant associated descriptions.[4] Where this thick form of descriptivism differs from the more standard forms which allow for existential antirealism is in its view that there is no theory-neutral description of the way the world really is, either in its deep structure or its observational features. Included among the terms which are defined by implicit definition there are not just the usual old theoretical terms but also any terms which an older philosophy of science used to call *observational.* As a result, goes the Kuhn-Feyerabend argument, no good sense can be made of the claim that certain terms belonging to a paradigm do not refer. Making sense of such a claim requires a notion of non-fit between paradigm and facts, and there is no interesting paradigm- or theory-neutral account of facts to be had. If our referential practices are aimed at a knowable and describable world, not a world of Kantian things-in-themselves, then we are bound to succeed in our reference with terms from our chosen paradigm — but so, of course, are those who use terms from radically different paradigms.

Those who forge this sort of link between the adoption of a thoroughgoing descriptivism and a relativist antirealism are generally quick to point to a way out for realists: give up descriptivism, and espouse instead the Kripke-(early)Putnam doctrine of *direct* reference, reference unmediated by concepts.[5] Nola's view is that some form of antirealism, either existential or relativist, is inevitable if descriptivism about the reference of scientific terms is accepted. If we wish to regain the real world, we need to drop descriptivism and replace it with something like the Kripke-Putnam doctrine of direct reference, with its austerely non-epistemic account of the way terms latch onto items in the world.

The Kuhn-Feyerabend version of antirealism is a relativist version of a kind of Kantian antirealism. Unlike existential antirealism, it doesn't deny existence but reconstrues existence as a kind of theory-dependent state. So relativist antirealism is a version of what Plantinga calls *creative* antirealism,[6] the view that objects somehow owe their existence and nature to the noetic conceptualizing activity of human beings.

Another much more general suggestion about the creative antirealist implications of descriptivism is found in Richard Rorty's interpretation of Kripke's work.[7] Kripke, according to Rorty, is strongly realist in his metaphysics, believing that the world is already divided up into particulars and natural kinds independently of any theory or conceptual scheme. Furthermore, this is not a world of Kantian things-in-themselves or of particulars and kinds that do not get a look-in until 10th milennium science. We can, to a large extent, name and describe the salient constituents of this world right now, whatever our favourite beliefs and theories, and consequently our terms must be able to hook onto these constituents independently of our beliefs and theories. The 'direct reference' approach to reference show how this is possible. Descriptivism, by contrast, is the theory of reference for creative anti-realists. It locates the identity of an object of reference in a speaker's way of thinking of, or conceiving, that object, and thereby encourages creative antirealism about such objects. It does so in a variety of ways. (What follows is my gloss on some of Rorty's remarks.) First, according to realism the world is independent of our beliefs. This is in part a modal point: objects might have existed without fitting our favoured descriptions of them. Descriptivism, however, encourages an *a priori* essentialism. If the word 'tiger', for example, means *large black-and-yellow striped feline*, etc., then tigers wouldn't exist unless there were large black-and-yellow striped felines. This suggests that to a degree tigers must be the way we think they are in order to exist, contrary to the demands of realism about objects like tigers.[8]

Secondly, for descriptivism there is in the end no *independently* accessible world that determines which of our terms refer. There are only a person's most settled beliefs involving that term, or perhaps the most settled beliefs of her fellow language-users. Thus let W be the way the world is represented as being. To claim that W is false, that W^* is the way the world really is, is to want to replace one set of descriptions by another. But on what principled grounds? Lacking the ability to refer in a description-free way to reality, about the only sense we can give to the claim that W^* is the, or a, best representation of the world is in terms of W^*'s being the, or a, most settled set of possible descriptions, that is, in terms of W^*'s satisfying certain internalist constraints rather than externalist constraints such as classical correspondence. In so far as this characterization does not invoke an interesting notion of an independently existing and independently accessible world but only our ability to come up with more or less coherent sets of descriptions, the metaphysical view it relies on is a version of creative antirealism.

Call the complex of considerations outlined the 'veil of descriptions' argument for obtaining (creative) antirealism from descriptivism. The existential, relativist and 'veil of descriptions' arguments are not the only arguments for linking descriptivism to antirealism,[9] but in one form or another they are probably the most familiar. The question now is how to evaluate them. Do these arguments really show that there is an antirealist thrust to descriptivism about reference, with different versions of descriptivism perhaps resulting in different versions of antirealism? It would, of course, be exciting if this were true. Realists would then not have a choice about which account of reference to favour. Only the choice of a theory of *direct* reference would show how reference to a world of mind-independent objects and kinds is possible.

Nonetheless I think all *DA* arguments are mistaken. I don't think that there is a valid argument from descriptivism, together, perhaps, with various independent background assumptions, to antirealism. I shall try to show this in a preliminary way by describing certain features which are *inessential* to descriptivism but which play a crucial role in generating antirealism via at least the existential and relativist *DA* arguments. Such a response, however, may be utterly effective on its own terms without really settling the more basic question of the relationship between the metaphysical doctrines of realism/antirealism and theories of reference. How can we know, in particular, that there does not exist a valid *DA* argument, one different from any of the disputed arguments? I propose to answer this question in a somewhat qualified but still reasonably strong form. I shall sketch an argument to show that a realist construal of the world, properly understood, best accords with a *descriptivist* account of how language-users manage to refer to the contents of this world. So if realism is on its terms a coherent doctrine, there can be no valid argument from descriptivism to antirealism.

2. REALISM

First things first, however. Without a more determinate understanding of realism, it is difficult to see what precisely is being claimed. This task is made especially difficult by the fact that realism has been understood in a variety of quite distinct ways, some of which carry absolutely no implications about semantic notions like truth or reference. One minimal version of realism — what Devitt calls *fig-leaf* realism — simply claims that there is something which exists in a suitably mind-independent way. Such a realism, however, is consistent with just about any ontology one might care for, including Kant's utterly unknowable and undescribable world of things-in-themselves. A 'fig-leaf' realism of this kind simply doesn't yield us a substantial enough external world, a world 'worth fighting for' (in Nelson Goodman's memorable phrase).[10] Such a world is 'an idle addition to idealism: antirealism with a fig-leaf' (Devitt, 1984, p. 15).

Another conception of realism that is currently much more in vogue is the one adopted by Dummett and his followers.[11] Realism in his sense involves the view that sentences from some designated class have verification- or evidence-transcendent truth conditions. From one point of view (but not from others) such a conception is again remarkably weak. It has virtually nothing to say about the contents and metaphysical status of the objective reality that makes sentences true, for even sentences about a world composed of sense-data may have verification-transcendent truth-conditions. Verification-transcendence has nothing to do with the nature of the *stuff* the sentences are about, only with the question of whether this stuff has its properties independently of our best criteria for telling what properties it has. From this point of view, then, but no doubt not from others, a realism in Dummett's sense doesn't yield us a world 'worth fighting for'.

The descriptivism-to-antirealism arguments we have looked at rested on a different and, I think, more intuitive conception of realism. This conception takes realism to be a substantial *metaphysical* doctrine, embodying both existence and independence claims concerning a range of entities of a certain broad type. I was able to concede the relevance of our three arguments to the realism/antirealism debate because in these arguments descriptivism was alleged to have the consequence that either crucial designators for objects or kinds *A* have no referent (thereby denying the *existence* dimension of realism about *A*'s) or what they refer to is somehow belief- or theory-dependent (thereby denying the *independence* dimension of realism about *A*'s).

More needs to be said, however. To begin with, what is being claimed to exist in realism *simpliciter* as opposed to realism about *A*'s? I am going to adopt, with some modifications to be spelled out, Michael Devitt's recent articulation of our intuitive conception of such a realism.[12] For Devitt, realism *simpliciter* is a certain kind of generalized version of particular realisms about *A*'s. Given a physical kind *A* (e.g. rock, tree, electron) let realism about *A*'s claim that 1) there are *A*'s and 2) these *A*'s have objective and mind-independent existence. Following Devitt, I shall mean by '*O* has objective existence' that *O* exists, and that its existence and nature are in no way dependent on our epistemic capacities (our knowledge, the synthesizing power of the understanding, the imposition of theories). Adding the point about mind-independence is supposed to ensure that real physical *A*'s are not constituted of objectively existing *mental* stuff (perhaps Kant's unsynthesized intuitions). Realism *simpliciter* can now be understood as the claim that for most current common-sense and scientific physical types *A* realism about *A*'s holds. So on this conception realism *simpliciter* is an amalgam of a common-sense and scientific entity-realism.[13]

There are questions aplenty to be asked about this characterization of entity-realism, not least the question of how best to characterize the dimensions of objectivity and mind-independence.[14] Let us not worry about these. I am here more interested in Devitt's claim that a realism of this sort (and *only* a realism of this sort) portrays a world 'worth fighting for'. But does it? Its commitments are greater than fig-leaf realism, of course, but are they sufficient to yield the world of common-sense and science, or, at any rate, enough of that world? The answer may seem self-evident. Realism claims explicitly that for most common-sense and scientific types of object *A*, *A*'s exist in the appropriate objective and mind-independent way. Note, however, that realism thus proclaims itself to be a view about existence. It says nothing about the metaphysical status of the usual sorts of ascriptions of properties to such entities. Consider, for example, the claim that trees produce their own food by synthesis, or that a particular tree has had its trunk charred by lightning. Realism on Devitt's conception is not committed to the view that facts such as these obtain objectively.

Now intuitively this is surely just as well. Realism should not commit itself to the obtaining of any particular set of facts. Indeed, for some the thought that the world may be quite different from the way we think it is is virtually constitutive of realism. Still, something is missing. When I said that realism on Devitt's conception is not committed to the view that 'facts such as these obtain objectively' I had in mind an even more extreme 'possibility'. Consider the following interpretation of Kant, one apparently contemplated by Schopenhauer and encouraged by Kant in the *Prolegomena*.[15] On this interpretation, Kant thought that things-in-themselves have two sides to them – their noumenal side (if you like, the set of *real* facts involving them) and their phenomenal side (the set of 'facts' involving them that are facts-for-us, i.e. for human beings, constituted as they are). If one has this view, then in talking of things as they really are and things as they appear to us one is talking of the same entities, but from different perspectives (from the possibly unoccupied God's-eye perspective and from the human perspective). What makes it true to say that, e.g., some particular tree has had its trunk charred by lightning is not that a certain noumenal fact obtains, but that a certain noumenal entity, cognitively presented to us as a tree, appears to humans to be a certain way under ideal perceptual conditions and conditions of rationality. That goes for any non-trivial truth about any object or kind humans are able to make reference to. The upshot of such a metaphysical view — call it existential noumenalism — is that the common-sense and scientific entities exist in an objective and mind-independent way without having any of their usual properties in an objective way. Even their most solid primary qualities are merely ways in which they appear to us in virtue of our cognitive constitution. I think that a world composed of such entities is not one that the realist ought to spend her time 'fighting for'. There is, in the end, no interesting difference between such a world and an antirealist world whose objects do not have objective or mind-independent existence but which display the same characteristics as the transcendently existing entities of the former.

Much of this picture borders on the unintelligible, no doubt (a usual complaint against things Kant said, or might have said about the noumenal). My aim is certainly not to defend the intelligibility of this doctrine, but to make the point that such a possibility for the world, if indeed it is any kind of possibility, should be ruled out *a priori* as a possibility for a worthwhile *realist* world. The question is how to do this. Some ways seem much too strong, e.g., urging that most of our beliefs about *A*'s better be *objectively* true if realism about *A*'s is to hold. Other ways are far too weak, e.g., urging that there are non-trivial facts about *A*'s which obtain objectively. Existential noumenalism certainly does not deny *that.* What it does deny is that such facts are expressible in our language, and that we can gain epistemic access to such facts. Both these propositions seem to me to constitute an important part of intuitive realism. If realism is to present us with a world 'worth fighting for', it ought to present us with a *knowable* world, or, at least, with a world much of whose real structure can be articulated in language and about which we can have many reliably-formed beliefs.

This suggests the following strategy. Let realism be given a small epistemic component E as follows: realism$_E$ about *A*'s claims that (1) there are *A*'s, (2) these *A*'s have objective and mind independent existence, and (E) (i) there are non-trivial, objectively obtaining facts about these *A*'s, both particular facts and facts about *A*'s in general, such that (ii) there are informational states available to humans which reliably certify these facts as obtaining.[16] 'Realism'$_E$ *simpliciter* is then defined from 'realism$_E$ about *A*'s' in the same way as 'realism' *simpliciter* is defined from 'realism about *A*'s.'

I shall call (E) the epistemic accessibility condition on realism$_E$. Note that this formulation of realism is still relatively weak. Realism$_E$ does not say that we can have epistemic access to *all* facts about *all* *A*'s with realist existence-conditions. No doubt there are many facts which we can never know about, perhaps because they are not expressible in language or because we cannot gain epistemic access to the entities in question. Although relatively weak, however, realism$_E$ still seems to advance claims that go well beyond realism's bare existential claims. One way of seeing this is to note that realism$_E$ has its own way of importing something like a semantic condition on realism. Realism$_E$ talks of the objective obtaining of humanly graspable facts, facts statable in language, and for those who don't like the apparent commitment to an ontology of facts implicit in this characterization it will seem imperative to use semantic ascent to talk of the *truth* of appropriate sentences. In addition, the truths involved are asserted by realism$_E$ to be sometimes knowable, or at least certifiable, by reliable means, as being true. Devitt strongly counsels against importing talk of truth into the characterization of realism ('realism says nothing about truth ... *Realism says nothing semantic at all*'[17]), but I think that his desire to have realism deliver a world 'worth fighting for' should have pushed him to embrace not only something akin to a condition involving truth but also something like the even stronger epistemic accessibility condition E[18].

3. DESCRIPTIVISM

So much for realism. Let us now get back to the issue we began with, the move from descriptivism about reference to metaphysical anti-realism licensed by our three *DA* arguments. The change to realism$_E$ doesn't change matters a bit. By the three *DA* arguments, we get failure either of the existence dimension or the objectivity dimension of a range of realisms (realism about *A*'s, for various physical types *A*), resulting perhaps in the falsity for existential reasons of scientific realism or, in the case of the relativist and 'veil of descriptions' arguments, in the wholesale falsity of realism. Realism, it seems, requires something different, an ability to refer in a way that is not mediated by descriptions, a theory of *direct* reference.

I think that the *DA* arguments, at least the first two, mistake their real target. They present a threat — how significant a threat is not important for my present purposes — to the conjunction of realism and what I shall call *classical* descriptivism. Classical descriptivism is what most philosophers have in mind when attacking, or even defending, descriptivism. It is a view that regards a speaker's conception of an object or kind as playing a multi-faceted role: the conception or description associated with a term

(1) is the purely conceptual, qualitative representation of an object which a fully competent speaker associates with her use of the term (this is a psychological role),
(2) determines via best fit what the term refers to (a semantic role),
(3) forms part of what is being *said* in the use of a sentence containing the term (a propositional role),
(4) gives the information-value of the term, and hence forms a part of any belief expressed by means of the term, thereby being relevant to the epistemological status — *a priori, a posteriori*, etc. — of the belief (a cognitive role).

In short, classical descriptivism supposes that the appeal to associated descriptions gives a rich notion of *meaning* for terms. Armed with such a view of the role of conceptions or descriptions, descriptivists don't have much choice about the content of reference-determining descriptions. Such descriptions are bound to be readily communicable, and to involve the ascription of salient, informationally vivid, criterially central, features: just the kind of descriptions favoured by Frege, Russell, and more recent cluster-theorists like Searle and Strawson[19] when characterizing the descriptive content of proper names. Importantly, they are not going to be egocentric in character, for that would flout the rule that they form part of what is being said or conveyed by embedding sentences. In saying 'John Smith was a famous whiskey maker', I am thus not saying, for example, that the person referred to by whoever informed me of the name 'John Smith' was a famous whiskey maker, for on that sort of view I won't ever say the same thing [utter the same proposition] as any other person who utters a token of that sentence — clearly a counterintuitive consequence.

Classical descriptivism thus characterized may well have antirealist consequences, as the three *DA* arguments assure us. But this joining of the fortunes of classical descriptivism and antirealism depends crucially on classical descriptivism's emphasis on the ascription of salient, informationally vivid, features. Both the existential and relativist arguments applied to the category of theoretical scientific terms, for example, will cite distinctive theoretical laws embedding the term when arguing that descriptivism leads to antirealism. What I disagree with in this is the thought that it is *descriptivism* which dictates the choice of such salient descriptions, and that it is therefore *descriptivism* which has the alleged antirealist consequences. This is a naturally drawn conclusion given the tendency to conflate the psychological, reference determining, propositional and cognitive roles descriptions may have, but we should resist it. Descriptivism, as I have stated and understood the doctrine, is fundamentally a theory of reference, not a theory of meaning. This is not to say that reference is *not* determined by meaning, in some rich sense of meaning in which it plays the various roles mentioned, but only that such a supposition would be an extremely strong theoretical addition to descriptivism as a theory of reference.[20]

Not surprisingly, divorcing the possible reference-determining role of descriptions from their other roles leads to a considerable widening of the field of descriptions available for a viable descriptivism about reference. What it achieves, in particular, is the re-admission of descriptions which are egocentric in character, and, more specifically still, descriptions which locate objects and kinds by reference to their causal/historical ties to term-users: 'the individual referred to by whoever informed me of the name "John Smith"', 'the red-looking celestial object I perceived yesterday', 'whatever causes such-and-such perceived effects'. Under classical descriptivism, such descriptions don't get a look-in. Under descriptivism, they do.

Merely making this point, however, doesn't invalidate the *DA* arguments. If all associated descriptions are to count, the salient, vivid ones will tend to dominate the comparatively insipid causal ones. What we might call *causal* descriptivism about the reference of a range of terms makes the further point that *only* broadly causal descriptions play a reference-determining role, where I shall leave the notion of a causal description suitably vague but mean it to include the ascription of causal powers and causal/historical connections to term-users, and exclude the sort of salient, distinctive descriptions appealed to in classical descriptivism. Indeed, the class of causal descriptions should include all those descriptions which theorists of direct reference generally import into their specification of the referential mechanism of proper names and natural kind terms (but which they do not, of course, think of as being used by referrers to play a self-conscious reference-determining role). Now we appear to have our objection to the *DA* arguments, at least to the existential and relativist ones: these two arguments are invalid, because they depend crucially on descriptivism being identified *ab initio* with classical descriptivism rather than with descriptivism *simpliciter*. An account like causal descriptivism is equally a version of descriptivism, but its emphasis on a relatively austere specification of causal relationships and causal powers provides an inhospitable environment for the existential and relativist conclusions drawn by the first two *DA* arguments.[21]

This still leaves the 'veil of descriptions' argument. That argument is the most general, and to that extent the hardest to evaluate, because it pretends to be able to do its job without relying on more specific information about the descriptions invoked. So it pretends to be able to do its job with causal as well as the salient descriptions emphasized by classical descriptivism. I shall leave this argument until the end of the paper where its claims can be more easily assessed.

Let me sum up briefly. I have tried to show that it is not descriptivism as such, but its standard classical form which arguably gives rise to antirealism via the *DA* arguments. Hence the *DA* arguments as arguments from *descriptivism* to antirealism are invalid. I want to consider just one objection to this response, but it is a fundamental one. Even causal descriptivism, it seems, gives rise to antirealism via at least the existential argument, since causal descriptions are risky in much the same way as ordinary descriptions ('the person I heard John refer to with *N*' is risky since you may be wrong about its having been John, etc.). If the appeal to inherently risky causal descriptions is ruled out, however (so the objection continues), the result is a contrived form of the doctrine which mimics in descriptivist terms what the theory of direct reference does much more directly, but achieves its success by appealing to abstruse descriptions that ought only to appear in someone's *theory* of how reference occurs.

One way of answering the objection might be to stress the narrow purpose served by our appeal to causal descriptivism. The claim which has been made is not that causal descriptivism is the most likely or the most natural form of descriptivism, but that because it is at least a *possible* version of descriptivism as such should not be seen as inherently antirealist in virtue of the *DA* arguments.

Maybe. But even so we will be left with some important loose ends. Perhaps the *DA* arguments can now be given as their focus philosophically intuitive or natural forms of descriptivism. This reformulation I would also reject, but such a rejection requires a better understanding of why the objection fails. More generally, showing that certain *DA* arguments fail doesn't show that descriptivism isn't inherently antirealist. There might be other valid arguments that show that even *causal* descriptivism is antirealist (as the 'veil of descriptions' argument presumably tries to do). One is tempted to ask for a consistency-proof for the compatibility of descriptivism and realism.

I propose to tie together these loose ends by sketching an argument for the strong conclusion already promised: if realism is true, it is descriptivism — in fact, a (largely) causal version of descriptivism — that best accounts for our ability to refer to the kinds of entities posited by realism. If this conclusion can be sustained, our other worries will have been shown baseless. There can, in that case, be no further worry about a 'veil of descriptions' which precludes reference to an objective, mind-independent reality and there can be no qualms about the 'unnaturalness' of causal descriptivism (unless, of course, realism itself is somehow seen to be the culprit, but that was never the intent of *DA* arguers).

4. THE ARGUMENT

Recall our earlier discussion of realism, and its defence of an epistemic component in realism. The claim I wish to argue for is that the theory of reference appropriate to the behaviour of referrers who succeed in referring to the kinds of objective, mind-independent item provided for by realism$_E$ is a version of descriptivism, largely a causal version of descriptivism. Perhaps the best way to argue for this claim is to take yet another detour through the Kantian heresy of existential noumenalism, but from a different point of view. What earlier motivated the reformulation of 'realism' as 'realism$_E$' was the somewhat vague *external* metaphysical concern that the world delivered by realism be 'worth fighting for'. Now consider the referrer's point of view, an internal perspective. A referrer who initiates a new referential practice or term-introduction does so because the initiation of such a practice is thought to have a point, and the initiation has a point if the selected object of reference (an individual or kind) is sufficiently salient for language-users to have an interest in discerning facts about the object (relational as well as intrinsic) and the practical ability to indulge this interest. Unless something like this is true, it is hard to see what the point of a new term-introduction would be. Frege's famous dictum that it is only in the context of a sentence that a name has a reference has had many readings, but one of the most plausible (on its own terms) is the claim that the purpose of naming is to secure reference to an object which we want to *talk* about, uncover *truths* about. If we fail to have any grasp on techniques that promise to yield us such truths, introducing a name lacks point. It is for this reason that mere causal connectedness is never sufficient to warrant a term-introduction. We introduce terms on the basis of the existence of what we expect to be a *cognitive* connection between objects and persons (paradigmatically, a perceptual connection), the kind of connection which allows a person to acquire information about the object to which he or she thus stands related.

I shall say more about this later, but right now I want to draw attention to the symmetry which exists between our (external) rejection of versions of realism that seemingly permit existential noumenalism concerning the entities realism is committed to, and the (internal) claim that the manner in which language-users initiate referential practices is constrained by the thought that they have a grasp of epistemic routes to the (alleged) objects of reference, a thought which rules out existential noumenalism concerning the objects of our reference.

The symmetry, I stress, is not manufactured by our illegitimately importing external considerations into our account of the internal dynamics of reference. I am not, for example, supposing that referrers themselves are implicitly realists$_E$, whether or not they implicitly acknowledge this. I am certainly inclined to believe this, but nothing much hinges on it. Even Kantian antirealists ought to accept the kind of account I have given of the internal dynamics of reference. If I am right, however, what Kantian antirealists ought *not* to say is that the entities we refer to are entities of which existential noumenalism is true. Allowing for the intelligibility of that doctrine, what Kantians ought to say is that we refer to things-as-they-appear-to-us, or appearances, rather than to the things which appear to us. Things-as-they-appear-to-us, like the epistemically accessible entities of realism$_E$, at least have the virtue of having their own set of knowable intrinsic and relational properties. What I claim is that it is the (expected) existence of an epistemic capacity to settle on a number of such an object's properties which, even on a Kantian account, ought to be seen as part of the assumptions constraining the initiation of a referential practice involving reference to the object.

In short, just as only the world of realism$_E$ is 'worth fighting for', so the only referential practices worth initiating from the point of view of language-users involve reference to objects which are self-consciously thought of as allowing certain types of epistemic access and hence as epistemically accessible. The rest of the argument now goes as follows. Let us say, slightly more formally, that the introduction of a term t has *epistemic warrant* for a person P just when (1) P believes that there exists something to be denoted by t, and (2) P believes that this object deserves denotation by a special term t in so far as (2a) she and other language-users have an interest in the object of a kind that, conditional on satisfaction of (2b), determines the desirability of a referential practice involving t, and (2b) she believes that she knows, to a sufficiently determinate degree, how one would set about determining facts about this object that are central to the kind of propositional enquiry appropriate to such an interest. Call (1) the *existence condition* on epistemic warrantedness, and (2b) the *fact-finding condition.* I have in effect, argued, no doubt all too briefly, that a term-introduction takes place only when it has epistemic warrant for the term-introducers, and hence only when the (intended) object of reference is conceptualized in a way that ensures the term-introduction's epistemic warrantedness.[22]

Generalities aside, however, how are the existence and fact-finding conditions satisfied in concrete cases? So far we only have a schema, no firm details. But I think that is just as well. What unifies all the various cases is the schema. There is no one way in which these conditions are satisfied, but a number of different ways. In the case of observable objects, substances and kinds, it is the thought that we stand linked to these items via perception, or the promise of perception, where perception allows for the cognitive state of observation and hence for certification of the item's existence as well as other information concerning the item. I do not rely here on a foundational notion of observation. The use of various kinds of microscopes, telescopes, computer-aided imaging, etc. should, I think, all count as observation, despite the enormous amount of theory that underlies the claim that these instruments allow us to acquire information about a source.[23] The point I am making about all such modes of observation is simply this: What warrants a term's introduction and the ensuing referential practice in the case of reference to an observable object or kind is not simply the existence of a direct causal link between observer and object, or even the existence of a perceptual link that in *fact* provides a reliable cognitive route to the object; it is, instead, the more or less explicitly represented thought that here we have an appropriate cognitive route to the object.

The case of theoretically-presented objects and kinds is in some ways more complex. The observational route is not available in such cases, so epistemic warrant must be generated in a different way. What generates epistemic warrant in this sort of case is the part of the theory that shows why we should assume the existence of the object, and how we might also reliably determine further truths about the object. In the case of causal-explanatory theories, this is very plausibly identified with the theory's specification of a particular causal-explanatory mechanism rather than its specification of a set of distinctive associated magnitudes or kind-constituting properties. Thus it is because phlogiston was thought of as a substance causally responsible for combustion and calcination by its characteristic emission from bodies undergoing these processes that phlogiston-theorists at once showed why they thought that phlogiston existed (answer: combustion and calcination need explaining; we have a powerful explanation) and how they thought further information about phlogiston could be acquired (answer: in conjunction with background 'knowledge', the appeal to a causal mechanism yields ready techniques for measuring and even isolating samples of the substance, for determining its chemistry, etc.).[24] It really does not seem enough for satisfaction of the fact-finding condition that phlogiston is austerely, but still causally, presented as *whatever* causes combustion and calcination. Therein lies no interesting research program concerning the causes of combustion and calcination. Stressing this point, of course, is going to be the way in which someone who wants to draw a descriptivist lesson from all of this will explain the fact that phlogiston isn't oxygen, that oxygen exists while phlogiston doesn't. *Nothing* displays the causal mechanism in terms of which phlogiston-theorists conceptualized the causes of combustion and calcination (all for the sake of epistemic warrant), and so phlogiston does not exist.

I *do* wish to draw a descriptivist lesson from these thoughts, both from the application of the idea of epistemic warrant to the case of theoretical scientific terms (where descriptivism might well be relatively uncontroversial) and from its application to the case of terms for observables. These applications encourage the thought that what determines an object or kind as the referent of a term introduced by a person or group is the fact that the latter thinks of it in a certain way *D*, where this way generates epistemic warrant for the term-introduction. *D*, furthermore, is typically a causal mode of presentation, hence the push for a descriptivism with a large causal descriptivist component. Thus *D* may involve the idea of something's being perceptually presented in a certain way (when an observable item is in question), or the idea of something's causing certain effects via the operation of a particular causal mechanism (as in the case of many theoretical scientific kinds). Various recalcitrant cases can be handled by involving more general types of explanatory mechanism.[25]

Still, this falls well short of *establishing* descriptivism. One concern centers on the sheer complexity of any viable account of reference-determination. Think of the multi-layered way in which a language-user's words stand semantically related to the object of her reference; theorists of direct reference like Devitt talk of comprehensive causal networks which underlie term-use (Devitt, 1981), and descriptivists need to be able to capture the complex of considerations that encourage this sort of picture.

Now I don't in fact believe that the complexity of our referential practices presents a deep problem. In particular, descriptivism can do at least as well here as a complex causal theory of direct term-reference. In Brian Loar's useful phrase, (Loar, 1976) causal descriptivism is simply 'the causal theory made self-conscious', and complicated versions of that theory can be made no less self-conscious than simple, wrong-headed versions. What my argument has tried to do is to sketch an intuitive motivational framework that shows why at the level of term-introduction referrers self-consciously conceptualize the objects of their reference in terms of the existence of cognitively useful causal relations. From the point of view of such an argument, complicating the pattern of relations will merely require a broader understanding of the interests of language-users, one which shows why we may realistically expect term-users to conceptualize the objects of their reference in terms of such more complicated patterns of relations.

There is a much more fundamental respect in which invoking the idea of epistemic warrant fails to establish descriptivism. Perhaps it is simply an interesting conceptual truth about the *point* of our referential practices that we are constrained to conceptualize the objects of our reference in this kind of way. Perhaps nothing follows from this about the conditions under which reference is successful. As Barry Stroud argues when commenting on transcendental arguments for the existence of an external world, how things really are may surely be quite different from the way we are somehow constrained to think they are, given our employment of certain concepts (Stroud, 1968). Whether or not Stroud is right about transcendental arguments, however, matters are surely different here. There is evidence aplenty to show that our interest in warranted referential practices has sharp consequences for the issue of the referential status of our terms, i.e., for whether or not they refer. Simply put, there is successful reference *just* when there are relevant bits of the world which conform to our epistemic-warrant-generating conceptualization. I think that the evidence for this is overwhelming (think about the cases of 'phlogiston' or 'Neptune'; or about what we would say if nothing fitted the causal-explanatory part of Yukawa's meson-theory, or we had been hallucinating rather than observing an exploding star when naming a new supernova; and so on), but I shall not rehearse that evidence here.[26] It all adds up to an impressive case for the descriptivist thesis that what determines the reference of terms, both ordinary names and natural kind terms as well as theoretical scientific terms, are (causal) descriptions, where the fact that descriptions play this role has a rather natural explanation in a kind of interest-driven account of the dynamics of our referential practices.

Let us see where we are in the overall argument. I have urged that descriptivism is the right view of reference, whatever things are like metaphysically. But this doesn't yet yield the strong conclusion that realism somehow accords best with descriptivism. We have to be sure that realism doesn't have in-built features that make the marriage of realism and descriptivism impossible. Before attempting to settle that issue, let me quickly get rid of one *bad* reason for supposing the marriage to be impossible.

On the version of descriptivism defended, the acknowledgement of a more or less definite epistemic route to an object of reference is implicit in the reference-determining conceptualization of the object. Surely, however, for the realist the entities of an objective, mind-independent world should not come packaged in this kind of way. Indeed, Devitt-style realism was characterized in terms of the *objective* existence of tokens of most common-sense and current scientific physical types, where to say that an object has objective existence is 'to say that its existence and nature is (*sic*) in no way dependent on our epistemic capacities' (Devitt, 1984 p. 13). Hence the suspicion that realism should be accompanied by an appropriately austere, non-epistemic account of how reference gets determined.

This line of argument concerns the theory of reference, note, not metaphysics. It doesn't attack realism$_E$, with its condition of epistemic accessibility, only the view that objects of reference come conceptualized as implicitly epistemically accessible. The argument is eminently resistable. The objectivity dimension of realism requires that the existence and nature of objects which are realistically construed are in no way dependent on our epistemic capacities. This rules out the view that the objects of our reference, where realistically construed, can have epistemic accessibility as one of their *essential* features. That way lies a strong version of anti-realism. But of course I did not take that route. Descriptivism, including causal descriptivism, is here regarded as a theory of reference, not a theory of meaning. So it doesn't treat the relevant causal-descriptive properties as *essential* features of what the term in question refers to, only as reference-determining features. Perhaps if one was a certain kind of antirealist one would have an argument for viewing them as essential, but that argument would be additional and would certainly not fall out of the theory of reference alone. Here, of course, we also have our reply to the first, modal, part of the 'veil of descriptions' argument. That part of the argument invoked the alleged antirealist consequences of an *a priori* essentialism that was supposed to be implicit in any version of descriptivism. The natural focus of such arguments is, I think not descriptivism about reference, but the sort of rich descriptivist account of meaning-cum-reference yielded by classical descriptivism.

We are just about done. The rest of the argument can now be played rather swiftly. Recall, once again, our weakly epistemic version of realism, realism$_E$, which became our preferred version because it was both relatively undemanding about the make-up of the real world and yet able to rule out, in an intuitive way, such unintended versions of realism as existential noumenalism. Realism can now be said to accord best with descriptivism for the following reasons:

(a) An epistemically motivated version of descriptivism is, in any case, the right view of reference, if our earlier argument was sound.
(b) Based on descriptions that implicitly present objects as allowing of practical epistemic access, this version of descriptivism does not

conflict with a realist conception of the objects of reference, and, in particular, with the objectivity dimension of realism. Indeed, realism is best characterized *ab initio* as a doctrine, realism$_E$, according to which the objects and kind-instances which have objective and mind-independent existence are to a degree epistemically accessible. Hence (c) realism accords well with descriptivism, and, given (a), accords best with descriptivism.

5. CONCLUSION

This argument, I have suggested, should allay our fears about *DA* arguments. I wish, finally, to make good on a promissory note: what, after all this, is the fate of the 'veil of descriptions' argument (the second non-modal part of that argument)? I offer the following suggestions. According to the descriptivism I have defended, there is no veil of descriptions in any relevant semantic sense. The descriptions that determine reference typically reflect our belief that we causally interact with the external world *via* certain types of cognitive pathways. If we are wrong about this, our terms will fail of reference. Otherwise, they will succeed. In addition, emphasizing an epistemic component in realism$_E$ enables us to see how even on realism our terms have a chance of succeeding at this task.

There seems to be nothing particularly mysterious here. How could there be? After all, one viable form of descriptivism, causal descriptivism, doesn't so much deny the existence of causal chains that lead into the referrer's head from external things (which, crudely, is what causal formulations of the theory of direct reference give us) as insist on putting the descriptions of these causal chains into the referrer's head, and letting the descriptions pick out the external things as referents. How can the theory of direct reference be more realist than a causal descriptivism of this kind, and how, therefore, can descriptivism be said to hide the world behind a 'veil of descriptions'?

On at least one understanding of the 'veil of descriptions' argument, however, this probably misses the point: Epistemologically, it might be suggested, there *is* a 'veil of descriptions': to evaluate sentences containing terms for external things, one has in effect to judge whether there is something meeting our conceptualization of these things, and there is no direct way of doing this — any candidate will itself come to us *via* a particular conceptualization. (This point is particularly obvious in the case of our reference to theoretical kinds, of course, since even 'observation' of the relevant candidates rests heavily on the use of background theory, so that our conceptualization of such entities as 'observed . . .' is going to have a heavy theoretical component.) The 'veil of descriptions' arguer now concludes that this shows the epistemic *inaccessibility* of a real world for the descriptivist. The only way to connect descriptivism and a commitment to a knowable world is to opt for antirealism and an epistemic notion of truth.

But why? Even the anti-descriptivist will not suppose that the kind of direct causal access to the real world talked of in theories of direct reference is somehow epistemically privileged access, providing conclusive grounds for existence-claims (and possibly other sorts of claims). Indeed, the thought that there is such access is itself arguably an antirealist thought.[27] Russell, of course, believed that there was such privileged access in the case of logically proper names, but he rightly saw that there were no such guarantees in the case of ordinary names, and so opted for descriptivism. Whether or not he was right to opt for descriptivism, he was surely right to argue *against* privileged access in this case. Clearly neither descriptivism nor the theory of direct reference works with a foundational sort of epistemology that reduces the gap between the world and our best epistemic methods to vanishing point. Where, then, is the relevant epistemological difference between (especially causal) descriptivism and the theory of direct reference? If anything, the shoe here is on the other foot. Theories of direct reference make the epistemological problem hard when they don't stress the *cognitive* nature of the causal word-world connections relevant to reference. The sort of descriptivism defended in this paper, on the other hand, is oriented directly towards the epistemological problem. Its claim is that descriptively understood referential practices are initiated on the basis of a reasonable (by one's own lights; how else?) expectation that the world contains certain objects and that it allows broadly specified kinds of epistemic access to these objects. Should our reasonable expectations be correct, we are able to gain epistemic access to the obects of our reference. If, furthermore, the world conforms to realism_E,we thereby gain epistemic access to parts of a *mind-independent* world. What more could a realist hope for from the theory of reference?

Perhaps I have still misunderstood the 'veil of descriptions' argument. Perhaps, in fact, the argument is really to be construed as an indictment of the conjunction of realism and *global* descriptivism, the view that the reference of all our terms is determined collectively by our global theory of the world or perhaps by highlighted (e.g., causal) segments of that theory. Only global descriptivism, it might be argued, is an all-out descriptivism covering *every* term of our language. Other versions of descriptivism are merely local, and presuppose that much of language already has determinate reference. If this is the way to understand the 'veil of descriptions' argument, I readily admit that I have not yet tried to provide an answer. For present purposes, I am content to admit that it is at most a (pervasive) local descriptivism that my argument has tried to show compatible with realism (indeed, to accord best with realism). Do matters change when we move to global descriptivism? Putnam thinks so, on the basis of his famous model-theoretic argument against realism (perhaps the 'veil of descriptions' argument *par excellence*). I think not, but that, as they say, is the subject of another paper.[28]

University of Auckland

NOTES

[1] Plantinga (1982, p. 48).
[2] Putnam (1976, pp. 183—4).
[3] See, for example, Kuhn (1970) and Feyerabend (1975).
[4] Devitt draws the descriptivism-antirealism inference in much the same way in his own account of the Kuhn-Feyerabend argument (Devitt, 1984, ch. 9).
[5] The theory of direct reference is more generally known as the *causal* theory of reference, although various philosophers sympathetic to the view dislike the emphasis on causality. Kim (1977) thinks that not all the external factors involved in reference may be reducible to causal factors, while even Kripke has expressed scepticism about whether the view he advances — Kripke (1972) is, of course, a *locus classicus* of the theory of direct reference — is in any crucial sense a *causal* theory (see Wettstein, 1986, fn. 13). Nathan Salmon is another mild sceptic, preferring to stress *contextual* elements in direct reference (Salmon, 1981). Perhaps the clearest and most elaborate *causal* version of the theory of direct reference is found in Devitt (1981). For reasons that should become clear in the paper, emphasis on the causal element seems to me to point in the right direction (although it will also become clear that I agree with some of Kim's reservations), and I shall generally have in mind *causal* formulations of the theory of direct reference.
[6] Plantinga (1982, p. 49).
[7] See especially Rorty (1980). Rorty (1982, pp. 122—3) is a different argument, more akin to the relativist *DA* argument.

[8] The example is from Devitt (1984, p. 20), but is in the spirit of Rorty (1980).
[9] Putnam (1980) and (1981) contain Putnam's model-theoretic argument for antirealism,where he takes (metaphysical) realism to be the view that even an empirically ideal theory of the world, as verified as can be, might be false of the actual world. [This is perhaps a form of the 'veil of descriptions' argument. Certainly Putnam believes that his own *internal* realism is a form of creative antirealism: 'the mind and the world jointly make up the mind and the world' (Putnam, 1981, p. xi).] And Devitt (1984) charges Dummett with an implicit reliance on descriptivism in his argument for a verificationist version of antirealism. Devitt maintains that acceptance of a causal account of reference blocks that argument by blocking the move from *understanding S* to *recognizing that the assertibility conditions for S hold.*
[10] Goodman (1978). Devitt uses the phrase with telling effect in criticizing various formulations of realism in Devitt (1984).
[11] See, for example, Dummett (1978).
[12] Devitt (1984, ch. 2).
[13] The quantification over physical types is eliminable in favour of a specification of the types in question (see Devitt, 1984, pp. 18—19). This would make it even clearer that we here have an *entity*-realism and not what might be called a *kind*-realism, the view that there are objective similarities in nature which make for natural kinds (a view Rorty, 1980, also attributes to Kripke). Like Devitt, I don't consider kind-realism here, and only consider the question: given the objective mind-independent existence of such things as tigers, rocks, electrons, etc. (an entity-realism), how can descriptivism about the reference of the terms 'tiger', 'rock', 'electron', etc. possibly be true? (A similar question could be asked about the consistency of descriptivism and a combined kind-and entity-realism. My answer would be much the same, but see note 28 below.)
[14] Hellman (1983) contains a good discussion of the problem of 'mind-independence' (which, for Hellman, is a combination of Devitt's 'objectivity' and 'mind-independence'), especially the problem of how to make sure that an apparently antirealist view like reductionist empiricism can't reconstruct the notion of mind-independence from within. Hellman also says good things in general about how best to define 'realism'. Like Devitt, he shows a preference for metaphysical over semantic formulations, with his preferred version of scientific realism claiming that 'much of science investigates a mind-independent material world.' From Devitt's point of view, however, this formulation still has problems. It seems consistent with science's *inability* ever to describe the world correctly or even to identify correctly any of the individuals or kinds it contains.
[15] See Kant's 'refutation' of the allegation of idealism in Kant (1950, Section 12). I owe this point of Kant-interpretation to Julian Young. (It turns out that there are strong traces of what I call existential noumenalism in the *Critique of Pure Reason* as well, especially in the first edition.)
[16] John Bigelow has pointed out to me that if one is a realist$_E$ about *A*'s then there are, or might well be, *B*'s such that one is *not* a realist$_E$ about *B*'s even though all *B*'s are *A*'s (take '*B*' to apply to all epistemically *inaccessible A*'s). Whatever else is worrying about this possibility, it seems to cause problems for my formulation of realism$_E$ because realism$_E$ can now fail to be true when the corresponding Devitt-style realism *is* true (because of such gerrymandered types *B*). I reject the charge, however. The formulation of realism and realism$_E$ was never supposed to work with such unnatural

descriptive types. Only the usual sorts of common nouns can replace '*A*' in the characterization of these doctrines, and it is the question of *their* referential functioning which is addressed by descriptivism and the theory of direct reference.

[17] Devitt (1984, p. 34).

[18] See also Taylor (1986) for a somewhat different defence of the claim that viable formulations of realism need to appeal to the objective obtaining of facts (or, for those who don't like facts, to the *truth* of sentences). It is important to note that the addition of an objective-obtaining-of-facts clause doesn't automatically turn such a characterization into the claim that particular *theories* (e.g. about electrons) state objectively obtaining facts or are broadly true. One can be a realist$_E$ about electrons and yet believe that the (knowable) truths about electrons exclude the deep theoretical statements found in typical electron-theories. In fact, one may even believe that *no* such theory can be true — the view, it seems, of Nancy Cartwright in Cartwright (1983).

[19] See Frege (1952), Russell (1956), Searle (1958) and Strawson (1959, ch. 6). Perhaps, however, it is no longer so very clear that Searle and Strawson are classical descriptivists. In this connection, see Searle (1982) and Strawson (1974, p. 59, esp. fn. 3). I am indebted to John Bigelow for the reference to Strawson.

[20] Burge (1977, p. 356) and Salmon (1981, pp. 12—14) make a similar point about how various distinct roles are conflated by the Fregean notion of sense. There are perhaps other roles which classical descriptivism conflates, but the present list makes the point well enough. Note that the propositional role is not clearly distinguished from the epistemic role by Burge or Salmon. I have done so because I am in fact fairly sympathetic to the objectualist view of propositions defended by Fine (1979), Almog (1986) and (to a degree) Wettstein (1986), so I am not inclined to assign a propositional role to descriptions associated with a name, although I *am* inclined to assign descriptions an epistemic role. According to Almog (1984) and (1986), this makes a study of reference-determination part of *pre*-semantics rather than semantics, which concerns itself with the structure of propositions. [I am somewhat dubious of this move, which strikes me as a piece of high redefinition. Lumsden (1987) is a good critique of Almog's reclassification.]

[21] This argument needs elaboration, of course, but I hope it is clear enough as it stands. Strip away the overt (but needless) commitment to a *non-descriptivist* causal theory of reference, and it is essentially the story told by Putnam (1975) (as well as in other writings by Putnam) and Nola (1980).

[22] See Kroon (1985) for more detail.

[23] See Hacking (1983, ch. 10) for an excellent account of the varieties of observation.

[24] The argument is given in Kroon (1985). This sort of account also serves to explain how a descriptivist can, in a non-arbitrary way, allow for referential commensurability in the case of theoretical terms occurring within widely varying theories *without invoking utterly austere (and unlikely) causal descriptions of the broad form 'whatever causes E in whichever way'.*

[25] Consider, for example, the way in which an appeal to group-theoretic considerations allowed Murray Gell-Man to predict the existence of particles of a given hypercharge, isotopic spin, electric charge and rest mass (the omega-minus baryons). The case is discussed by Freeman Dyson in Dyson (1964). Examples of this type accord with our stress on epistemic warrant, and indicate the degree to which causality is somewhat of a

red herring in the theory of reference (although causality is also important; *why* it is important also falls out of a view stressing epistemic warrant, given the epistemic centrality of causal explanations and of the idea of observational access).

[26] That is, we ask, concerning other possible worlds, how things would be referentially if the epistemic-warrant-generating descriptions fitted nothing. I claim that the answer is clear: the terms would be referentially empty. This is an application of a kind of semantical argument which [as Salmon (1981, pp. 29—30), points out] is used to great effect by Kripke and Putnam to dispute classical descriptivism.

[27] See Papineau (1985) for an argument that the commitment to belief-generating methods which cannot help but be rational is an antirealist position.

[28] Not necessarily my own. See Merrill (1980) and, following Merrill, Lewis (1984) for a certain kind of realist response to Putnam's argument. The realist response in question invokes what in note 13 I termed a *kind-realism*, the thought that the world divides into kinds not of our own making, that there are objective joints in nature. These will then constrain the possible 'intended' interpretations of our ideal theory, falsifying Putnam's claim that any (empirically ideal) consistent theory will have many intended interpretations if the cardinality of the world is large enough to offer us *one* interpretation on which the theory comes out true. (As Merrill, 1980, fn. 11, points out, Putnam himself once held such a kind-realism.) It should also be noted (as Lewis, 1984, pp. 231—232, does) that the model-theoretic argument at most shows that there are too many interpretations on which the theory comes out true. It does *not* show that there is not a world of objectively existing, mind-independent entities, only that there are too many ways of moulding these into 'true-making' interpretations.

REFERENCES

Almog, J. (1984) 'Semantical Anthropology', in P. French, T. Vehling, and H. Wettstein (eds.), *Midwest Studies in Philosophy*, vol. 9, Minneapolis: University of Minnesota Press, 479—490.

Almog, J. (1986) 'Naming without Necessity', *The Journal of Philosophy* **83**, 210—242.

Burge, T. (1977) 'Belief *De Re*', *The Journal of Philosophy* **74**, 263—282.

Cartwright, N. (1983) *How the Laws of Physics Lie*, Oxford: Clarendon Press.

Devitt, M. (1981) *Designation*, New York: Columbia University Press.

Devitt, M. (1984) *Realism and Truth*, Oxford: Basil Blackwell.

Dummett, M. (1978) *Truth and Other Enigmas*, Cambridge: Harvard University Press.

Dyson, F. J. (1964) 'Mathematics in the Physical Sciences', *Scientific American*.

Feyerabend, P. (1975) *Against Method*, London: New Left Books.

Fine, K. (1979) 'First-Order Modal Theories II-Propositions', *Studia Logica* **39**, 159—202.

Frege, G. (1952) 'On Sense and Reference' (originally published in 1892), P. Geach and M. Black (eds.), *Translations from the Philosophical Writings of Gottlob Frege*, Oxford: Basil Blackwell, 56—78.

Goodman, N. (1978) *Ways of Worldmaking*, Indiannapolis: Hackett.

Hacking, I. (1984) *Representing and Intervening*, Cambridge: Cambridge University Press.

Hellman, G. (1983) 'Realist Principles', *Philosophy of Science* **50**, 227—249.

Kant, I. (1950) *Prolegomena to Any Future Metaphysics* (originally published in 1783). L. W. Beck (ed. and transl.), New York: Bobs-Merrill, Library of Liberal Arts.

Kim, J. (1977) 'Perception and Reference without Causality', *The Journal of Philosophy* **74**, 606—620.

Kripke, S. (1972) 'Naming and Necessity', in Davidson D. and Harman G. (eds.), *Semantics of Natural Language*, Dordrecht: Reidel, 253—355.

Kroon, F. W. (1985) 'Theoretical Terms and the Causal View of Reference', *Australasian Journal of Philosophy* **63**, 143—166.

Kuhn, T. (1970) *The Structure of Scientific Revolutions*, 2nd edition, Chicago: Chicago University Press.

Lewis, D. (1984) 'Putnam's Paradox', *Australasian Journal of Philosophy* **62**, 221—236.

Loar, B. (1976) 'The Semantics of Singular Terms', *Philosophical Studies* **30**, 353—377.

Lumsden, D. (1987) 'What's in (the Semantics of) a Name?', Paper delivered to the New Zealand Division of the *Australasian Association of Philosophy*, Wellington.

Merrill, G. H. (1980) 'The Model-Theoretic Argument Against Realism', *Philosophy of Science* **47**, 69—81.

Nola, R. (1980) '"Paradigms Lost, or the World Regained" — An Excursion into Realism and Idealism in Science', *Synthese* **45**, 317—350.

Papineau, D. (1985) 'Realism and Epistemology', *Mind* **94**, 367—388.

Plantinga, A. (1982) 'How to be an Anti-Realist', *Proceedings of the American Philosophical Association*, 47—70.

Putnam, H. (1975) 'Language and Reality', in H. Putnam, *Mind, Language and Reality: Philosophical Papers, vol. 2*, Cambridge: Cambridge University Press.

Putnam, H. (1976) 'What is "Realism"?', *Proceedings of the Aristotelian Society* **76**, 177—194.

Putnam, H. (1980) 'Models and Reality', *Journal of Symbolic Logic* **45**, 464—482.

Putnam, H. (1981) *Reason, Truth and History*, Cambridge: Cambridge University Press.

Rorty, R. (1980) 'Kripke versus Kant', *London Review of Books*, 4 Sept.—17 Sept., 4—5.

Rorty, R. (1982) *Consequences of Pragmatism*, Minneapolis: University of Minnesota Press.

Russell, B. (1956) 'The Philosophy of Logical Atomism' in R. Marsh (ed.), *Logic and Knowledge*, London: Allen and Unwin.

Salmon, N. V. (1981) *Reference and Essence*, Princeton: Princeton University Press.

Searle, J. (1958) 'Proper Names', *Mind* **67**, 166—173.

Searle, J. (1982) 'Proper Names and Intentionality', *Pacific Philosophical Quarterly* **63**, 205—225.

Strawson, P. F. (1959) *Individuals: an Essay in Descriptive Metaphysics*, London: Methuen.

Strawson, P. F. (1974) *Subject and Predicate in Logic and Grammar*, London: Methuen.

Stroud, B. (1968) 'Transcendental Arguments', *The Journal of Philosophy* **65**, 241–256.

Taylor, B. (1986) 'The Truth in Realism', preprint, Department of Philosophy, University of Melbourne.

Wettstein, H. (1986) 'Has Semantics Rested on a Mistake?', *The Journal of Philosophy* **83**, 185–209.

GRAHAM ODDIE

ON A DOGMA CONCERNING REALISM AND INCOMMENSURABILITY

It is widely held that realists must accept a strong version of the commensurability of theories, and anti-realists a strong version of the incommensurability of theories. It is the purpose of this paper to throw doubt on this dogma, a dogma which seems to be shared by realists and anti-realists alike. In the first part of the paper I will show that one extremely influential line of reasoning for the connection between anti-realism and incommensurability (perhaps the only line of reasoning) is mistaken. It will emerge that anti-realists need not eschew commensurability in a large class of cases often thought to be prime candidates. In the second part of the paper I will show that realists are committed to theses which drive them rather far down the slope of incommensurability.

1. REALISM

A fundamental tenet of realism, call it the *truth doctrine*, is that the aim of an inquiry is the truth about the structure of the world. An inquirer need not seek the whole truth about *every* aspect of the world. Each inquiry delimits an *aspect* of the whole truth (for example, the truth about the number of planets, or about the laws governing the planets, or about the immortality of the soul, or whatever) but the aim of the inquiry, *qua* inquiry, is to get at the whole truth about that particular aspect of the world's structure.

Another tenet of robust realism is that an inquiry may seek an aspect of the truth which is, strictly speaking, beyond complete empirical decidability, even in principle. Empirically adequate theories may nevertheless be false. This will be called the *verification transcendence doctrine*.[1]

A third tenet of realism, not widely appreciated as such, is that it is at least possible for an inquiry to make (non-trivial) progress in its aim, without actually achieving it completely. That is to say, of two theories one may be a better approximation to the truth than the other, despite

Robert Nola (ed.), Relativism and Realism in Science, 169–203.

the fact that both fall short of the whole truth of the matter in question. This will be called the *possibility of progress* doctrine.

There may be more to realism than these three doctrines, but one view which it is implausible to impute to realism is the view that our current inquiries, scientific or otherwise, have already attained their goal, or even that they represent progress over previous stages of the inquiry. Realism will not have been refuted, or even damaged, if current theories turn out to be wildly wrong, or more wrong than their predecessors. Realism is a philosophical view about the nature of inquiry generally, and the relation of an inquiry to a world which is in some sense inquiry-independent. It is not a contingent thesis about the successes or failures of actual inquiries, and it is entirely compatible with robust fallibilism.

While there may be more to realism than these three doctines, any view which eschews one or any combination of these is either a severely impoverished realism, hardly worthy of the title, or not realism at all. A brief look at the possibilities may make this claim plausible.

There is no logical difficulty in espousing doctrines one and three and eschewing doctrine two; but the result usually goes under the name of *internal realism*, to mark its distance from robust, full-bodied, metaphysical, or external realism.

Again, doctrines one and two may be espoused while doctrine three is rejected, but the resulting position would be profoundly depressing for most realists. If the only theory which is any good is the one which fulfils the aim of inquiry completely then every miss is as good as a mile. There can be no gradual progress in an inquiry (*qua* inquiry). Either the whole truth is hit upon, or the enterprise is an utter failure. Even a *truth* which gives only part of the truth of some matter would have to be considered worthless, because it would not realise the aim completely. A realism which eschewed the possibility of progress would be forced to accept an extreme pessimism not only about the course of actual inquiries, but of all possible inquiries, excluding only the most trivial. It is only by accepting the possibility of progress that the realist's aim can seem anything other than hopelessly utopian.

Both the second and third doctrines presuppose the first, and a rejection of the truth doctrine would entail a rejection of all three. Needless to say, any such rejection would be totally alien to any kind of realism.

It is worth noting here that the doctrines of truth and of the possibility of progress demand a notion of closeness to truth (or truthlikeness, verisimilitude). For if progress towards the truth is to be possible then it must make sense to say of two theories both of which fall short of the whole truth, that one is closer to the truth (or more truthlike) than the other. Popper was, of course, the first philosopher to stress the importance of the concept of truthlikeness, but having a coherent account of the concept is not merely a requirement of a Popperian account of knowledge. It is a general requirement of realism. The reason Popper was the first to stress its importance is simply that Popper was the first philosopher to take seriously two theses: firstly, that our current theories, like their predecessors, are (almost certainly) false; secondly, that our current theories *may* nevertheless realise the aim of inquiry better than did their predecessors. Popper's concern for truthlikeness thus sprang out of his combination of fallibilism and optimism. But strictly speaking, all that is required to generate the concern are the two rather modest doctrines of truth and the possibility of progress.[2]

2. INCOMMENSURABILITY

The thesis of incommensurability is vaguer than that of realism, but if we take, more or less at random, three different characterisations given in the literature, then a few core tenets emerge.

Devitt (in 'Against Incommensurability') writes that the incommensurability thesis is:

> ... the thesis that different theories in the one area may be radically incomparable because of meaning changes.[3]

Musgrave (in 'How to Avoid Incommensurability'):

> The incommensurability thesis says that successive major scientific theories, or paradigms, or world-views, are incommensurable because the meanings of terms occurring in them are different.[4]

English (in 'Partial Interpretation and Meaning Change'):

> Because the meanings of their terms change, we are told, competing theories are expressed in different, not readily intertranslatable, languages. Rival theorists tend to "talk past" each other: their arguments "fail to make contact"; it is as though they were members of different language-culture communities. It is in this sense that rival theories may be said to be "incommensurable".[5]

Thus incommensurability seems to amount to incomparability due to changes in meaning of crucial terms. But comparability is always relative to some feature or aim, and it is never made very explicit exactly what feature or aim is at issue here. Discussions almost automatically slide into the controversy over meaning changes before it is settled what kind of incomparability such meaning changes ensure. But if, as is usually assumed, incommensurability is a problem for the realist, and we take as our characterisation of realism the doctrines outlined in Section 1, then it is easy to pinpoint the aim or feature with respect to which theories are claimed not to be comparable: incommensurability is incomparability with respect to the aim of inquiry. That is to say, two theories are incommensurable if they cannot be compared for the degree to which they succeed in capturing the truth. Or what amounts to the same thing, they cannot be compared for truthlikeness.

There are two distinct, but connected, senses of 'comparable' here: a purely logical sense, and an epistemological sense, and there are two distinct, but connected, problems of truthlikeness. The *logical* problem of truthlikeness is to specify what it takes for one theory to be closer to the truth than another. The *epistemological* problem is to specify what would constitute evidence for a judgement of relative truthlikeness. The logical problem has priority, in that a solution to the epistemological problem presupposes a solution to the logical problem. Consequently there are two kinds of incommensurability. Two theories are epistemologically (or weakly) incommensurable if there are no evidential criteria for judgements of the relative truthlikeness of the two theories. They are logically (or strongly) incommensurable if judgements of their relative truthlikeness simply do not make sense. As logical incommensurability entails epistemological incommensurability, and it seems to be what most proponents of incommensurability have had in mind, it is the kind of incommensurability discussed here.

Other aspects of incomparability have been canvassed but it is not implausible to regard these as important only because of their indirect links, or purported links, with truthlikeness. Consider just two such aspects: the lack of apparent logical conflict between incommensurable theories and incomparability for content. One simple but appealing kind of theory of truthlikeness is based on the idea that aggregation of truths and removal of falsehoods constitutes progress towards the truth. Without too much reflection on the issue this may well be the most natural kind of theory for a philosopher to start working on. Although

Popper was the first to try to articulate this theory in a precise form, it is nevertheless probably true that the theory is lurking in the back of most people's minds.[6] If this is right then it is clear why, for example, logical conflict is important for truthlikeness. One important way forward would be to root out a falsehood and replace it with its negation. But if apparently rival theories 'talk past each other' and apparently contradictory statements are not really so, then this simple model cannot be applied.

Or consider content. One way of articulating the crude theory adverted to above is to compare the truth content and falsity content of two theories. The truth content is the set of truths a theory implies, the falsity content the set of falsehoods it implies. Popper's first theory of truthlikeness, and it is very tempting, claims that either a set-theoretic increase in truth content, or a set-theoretic decrease in falsity content, guarantees a step towards the truth. But if the contents of theories cannot be compared (because the same sentences have very different meanings) then obviously this idea cannot be applied.[7]

We have then a *definition* of incommensurability. The *thesis* of incommensurability can be split into two parts. The first is a purely conceptual claim to the effect that meaning variance guarantees incommensurability. The second is a factual claim to the effect that apparent rivals in the history of inquiries are incommensurable because of such meaning variance. Most of the analysis that ensues will be concerned with the conceptual claim, but the conclusions will have relevance for the factual claim as well.

3. ANTI-REALISM AND COMMENSURABILITY

For our starting point we may take a brief, but characteristically bold, article by Alan Musgrave (1979): 'How to Avoid Incommensurability'. In it Musgrave points to the slippery slope from incommensurability to anti-realism [see Nola (1980) for further details] and then argues that there is no good reason to get onto the slippery slope in the first place. This is because there is only one decent argument for the incommensurability thesis, and the major premise of that argument is false. This master argument begins with the doctrine that a theory partially defines its own theoretical terms (the doctrine of partial interpretation). Granted this doctrine it seems that the meanings of theoretical terms will be theory dependent. Substantial differences in theory will ensure substantial differences in meaning, and incommensurability seems to follow. Musgrave argues that the doctrine of partial interpretation is fundamentally mistaken, and that this emerges from the failure of the various attempts to articulate it rigorously.

Musgrave is right, I think, to stress that the main inspiration for meaning-variance theories is the idea that theories partially define or interpret their own theoretical terms. But there is available one very attractive account of partial interpretation which is not susceptible to Musgrave's criticisms: namely, Carnap's last proposal.[8] What is not widely realised, and will be argued below, is that Carnap's proposal does not generally entail meaning-variance for rival theories.[9] Moreover, the proposal is compatible with a very large measure of commensurability amongst theories, even in those cases in which meaning variance does arise. Thus a consideration of Carnap's proposal shows not only that the arguments from partial interpretation to meaning-variance, and from meaning-variance to incommensurability, are invalid. Because it is, in at least one important sense, anti-realist, it also furnishes an example of one brand of anti-realism which violates the dogma of the supposed link between anti-realism and incommensurability.

According to the positivist tradition, out of which Carnap's proposal grew, the meaning of theoretical terms is problematic. The positivist programme was to show how such terms could be invested with meaning by being connected with observational terms. Without such a connection statements incorporating the terms would be metaphysics, or meaningless, or both. Carnap in particular was deeply concerned with this problem of the meaning of the theoretical terms, and made a series of proposals each of which was more holistic and less restrictive than its predecessor. In his early work he demanded that theoretical terms be explicitly definable in observational terms. His last proposal incorporates Quine's insights to the effect that theoretical terms gain their significance only by virtue of their role in a theory, or whole cluster of theories, a cluster which only confronts experience at the borders; and that even then the meanings of such terms are underdetermined. However, Carnap rejected Quine's pessimism as to the possibility of separating off the stipulative from the factual content of a theory. And his last proposal demonstrates the compatibility of Quinean holism with the thesis that analytic and synthetic components of a theory are distinct.

As is well known, Ramsey gave a general method for separating off the observational content of any (finitely axiomatisable) theory or system of theories.[10] Let V_O be a collection of observational terms (however these are to be characterised) and V_T a collection, $\{T_1, \ldots, T_m\}$ of non-observational terms, and let A be a finite axiomatisation of some theory, along with whatever else a Quinean holist might want to include (low-level observational theories, correspondence rules connecting theoretical and observational terms, and so on). Let $A(t_1/T_1, \ldots, t_m/T_m)$ be the open sentence obtained from A by replacing each theoretical term T_i with a variable t_i ranging over the sort of entity T_i is supposed to denote. Then the Ramsey sentence of A, A^R, is the existential closure of $A(t_1/T_1, \ldots, t_m/T_m)$:

$$(A^R) \quad (\exists t_1) \ldots (\exists t_m) A(t_1/T_1, \ldots, t_m/T_m).$$

Ramsey proved that for any sentence S_O which contains only observational terms, S_O follows from A if and only if S_O follows from A^R.

$$\text{(Ramsey)} \quad A \models S_O \quad \text{if and only if} \quad A^R \models S_O.$$

That is to say, A^R gives us all the observational content of A. To say that A is observationally adequate (or empirically adequate) is just to say that every observational consequence of A is true. And Ramsey showed that this is tantamount to the truth of A^R.

Being a positivist, Carnap was happy to identify the factual (synthetic) content of A with its observational content, A^R. Thus Ramsey had done almost half the work in splitting up factual and stipulative components of A. The problem remained of isolating the analytic or stipulative component of A. Two desiderata immediately suggest themselves. Firstly, the analytic component must not entail any non-trivial observational consequence: it must be *observationally uncreative*. Secondly, the analytic and synthetic components together must be formally tantamount to the original theory. These two desiderata are satisfied by what is sometimes called the *Carnap sentence*, A^C:

$$(A^C) \quad A^R \supset A.$$

Carnap's proposal is that A^C be taken to be the analytic component of A, and A^R the synthetic component. There are other sentences satisfying the two desiderata. For example, if B is any sentence over V_T (a purely theoretical sentence) which A entails then the conjunction of

A^C and B also satisfies the desiderata. In fact A^C is, formally, the 'weakest' statement to satisfy the desiderata, and Carnap does not give any convincing justification for preferring it to other stronger statements. Fortunately, Winnie has supplied just such a justification. Winnie shows that the Carnap sentence is the strongest sentence of the theory satisfying the condition of *observational vacuity*.[11] A sentence satisfies the condition just in case it plays an inessential role in generating observational sentences from the theory, even in combination with other sentences of the theory. Observational vacuity entails the first desideratum, but it is stronger, and in conjunction with the second desideratum, yields A^C as the only candidate for the analytic component.

Note that Carnap's proposal does not suffer from the defect Musgrave criticises in other accounts of partial interpretation: namely that the theoretical postulates are rendered true by terminological fiat.[12] In fact *none* of the purely theoretical postulates of A turns out to be analytic on this account (provided of course, that they are not logically true).

The semantic justification for Carnap's proposal is interesting. It is an admirably clear version of Putnam's much discussed model-theoretic argument against realism, and it was effectively available a decade before the latter. What can be shown, almost immediately, is that if A is observationally adequate then there is a way of interpreting the theoretical terms of A in such a way as to render A true. For suppose A is observationally adequate, that is A^R is true under the interpretation I_O of the observational vocabulary V_O. Since A^R is $(\exists t_1) \ldots (\exists t_m)A(t_1/T_1, \ldots, t_m/T_m)$ it follows, from the usual Tarskian rules, that there is an assignment of sets $S_1, \ldots, S_m$ to the variables $t_1, \ldots, t_m$, which satisfies the open sentence $A(t_1/T_1, \ldots, t_m/T_m)$. Now consider the extension I of I_O which assigns S_i to T_i. I is an interpretation of the full vocabulary V and, again by the usual Tarskian rules, A must be true in I. Assuming for the moment that to assign an extension to a term is to interpret it, this shows that the observational adequacy of a theory guarantees the existence of interpretations which render the theory true. To stipulate, with Carnap, that A^C be analytic is to rule out as inadmissible other interpretations of the theoretical terms. It is to rule that whenever a theory is observationally adequate then its theoretical terms are to be understood in one of the ways which renders the theory true. Alternatively, if implication and equivalence are defined not in terms of *all* interpretations, but in terms of *admissible* interpretations, then Carnap's proposal is tantamount to taking the theory to be equivalent to its Ramsey sentence. In all admissible interpretations, A and A^R have the same truth value.

Carnap had no compelling argument that theoretical terms *must* be so interpreted. Indeed, given that the proposal is tantamount to the positivist doctrine that the factual content of a theory is identical to its observational content it is hard to see how any argument for the proposal could command support generally. Somewhere in the premises there would have to be a bias towards positivism.

It is clear in what sense a theory partially defines its theoretical terms. Although taking the Carnap sentence to be analytic restricts interpretations to those that are admissible, it will not usually single out a unique interpretation, or even a unique interpretation for each interpretation of V_O. Generally there will be a range of admissable interpretations of the theoretical terms. A partial interpretation is thus a specification of a *range* of completely determinate interpretations, each of which is equally legitimate from the point of view of the theory.

As I have set it out here Carnap's proposal is couched within an extensionalist theory of interpretation, according to which an interpretation of a vocabulary is an assignment of extensions (sets to predicates, truth values to sentences). While this kind of approach may be perfectly acceptable for mathematical languages (for which it was explicitly designed) it has obvious and well-documented shortcomings for factual languages. It is strange that Carnap should not have been sensitive to this given that he himself laid the foundations for an adequate intensionalist account of interpretation, in terms of possible worlds, in his (1947). Many philosophers tend to think of worlds and (extensional) interpretations as interchangeable, and in this they have been encouraged somewhat by Carnap's early work. However, Carnap himself was extremely critical of this conflation in his (1971),[13] although he did not spell out in detail the kinds of confusions it engenders. Briefly, an interpretation should assign to a (factual) predicate a property, and this property does not change its *identity* in different possible worlds, though it may well have different *extensions* in different possible worlds. A world is not an assignment of either sets or properties to syntactic items, but is a way things might be, or might have been. As such it can be explicated as an assignment of extensions to traits or properties, but these are non-syntactic items.

When one considers higher-order systems and languages the distinction is absolutely crucial, because in these cases (extensional) interpretations cannot ape the features worlds have. But even in the first-order case it is important to make the distinction, in order, for example, to frame an adequate account of the structure of a state of affairs.[14] In any case, is interesting to investigate whether or not the Ramsey-Carnap methods can be extended to an intensionalist theory of interpretation, and this is done below. It is shown that the methods can be extended but only under certain restrictive conditions. But my main purpose here is not to argue that the intensionalist approach is the correct one: rather, it is to show that whether one accepts an intensionalist or an extensionalist approach, Carnap's proposal allows a very large degree of meaning compatibility amongst rival theories. As such, Musgrave's criticism of it, that it does not allow rival theories of, for example, electrons, seems to be unfounded.

Recall that according to the proposal a theory A yields a partial interpretation of the theoretical terms, and that a partial interpretation is a range of completely determinate interpretations, any one of which is equally legitimate from the point of view of the theory. It is then natural to say that A and B are *meaning compatible* just in case there is at least one way of interpreting V_T which is compatible with the partial specifications of both A and B.[15] That is to say, A and B are meaning compatible just in case they do not ensure that their common theoretical terms have different meanings. The conditions for meaning compatibility will depend on the theory of interpretation one adopts.

Firstly, consider an extensionalist account of interpretation. A cuts down the set of such interpretations to those in which A^C is true: the A-admissible interpretations. And a sentence is analytic if it is true in all A-admissible interpretations. Obviously the sentence $A \supset A^R$ is analytic, so that A is equivalent to its Ramsey sentence. A and B will be said to be *weakly* meaning compatible just in case there is one interpretation which is A-admissible and B-admissible. We will introduce a stronger notion below in response to an objection to weak meaning compatibility.

PROPOSITION 1. *A and B fail to be weakly meaning compatible if and only if they are logically incompatible and both are observationally uncreative.*

(For the proof, see the appendix.) From this result it is clear how extraordinary meaning incompatibility is. For A to be meaning incompatible with B, A and B must contradict each other but neither must have any non-trivial observational consequences. From the positivist point of view this is an extremely desirable result. It demonstrates that two *purely* metaphysical theories which *appear* to be in conflict are really just 'talking past each other': there is no way of giving a common interpretation to their theoretical vocabulary. But as a matter of course, any two theories with some non-trivial observational content, and this includes observational rivals, will be meaning compatible. Rival theories of electrons are possible, provided at least one of them has a non-trivial observational consequence.

The philosopher committed to meaning variance might balk at this definition of (weak) meaning compatibility. She might insist that some of the interpretations which guarantee compatibility would be bizarre. It might assign to observational terms extensions which are not their actual 'intended' interpretations. (The same cannot be said about the theoretical terms without abandoning the idea that it is the theory alone which delimits the legitimate interpretations of these.) The objection is a difficult one for an extensionalist to press, because it quickly brings to light the implausibility of the basic theory. It is strange indeed to say that the actual set of green things is the *intended* interpretation of the predicate 'green' simply because in order to use and understand the term perfectly it is hardly required that the speaker know *which* set that is. And since speakers do not know which set that is, it is peculiar to insist that that is the set they *intend* to pick out. It may be tempting to defend the theory by pointing out that although the speaker does not know the exact identity of the extension of the set he intends (and since sets *are* extensional, he does not know *which* set he intends) he does intend to pick it out *as* the set falling under a certain description ('the set of green things'). But of course this is to simply cast the very same problem in a different guise, for we are now faced with rival extensional and intensional accounts of the meaning of definite descriptions.

In any case, let us grant to the objector the thesis that the intended interpretation of the observational vocabulary V_O is an assignment of extensions to the terms in V_O, say I_O. In this case it would be natural for the extensionalist to restrict admissible interpretations to those that are extensions of I_O. An interpretation of V is *strongly A-admissible* just in case it is an extension of I_O and A^C is true in it. And A and B are *strongly meaning compatible* just in case there is at least one interpretation which is both strongly A-admissible and strongly B-admissible.

PROPOSITION 2. *A and B fail to be strongly meaning compatible if and only if A and B are observationally adequate while A & B is not observationally adequate.*

(Again, the proof is in the appendix.) Although it is more difficult to be strongly meaning compatible than it is to be weakly meaning it is still rather easy. To fail to be meaning compatible both theories have to be individually observationally adequate while their conjunction fails to be so. Again observational *rivals* (that is theories which, in conjunction with known or assumed observational facts, deliver incompatible observational predictions) will be meaning compatible. Rival theories of electrons are thus possible.

In order to extend the Ramsey-Carnap methods to a full-blown intensionalist theory of interpretation it is necessary to spell out in some detail how such a theory is to be constructed. This is done in the appendix for a class of higher-order languages ideographically very similar to first-order languages. Briefly, an interpretation of a language assigns to symbols appropriate objects over a logical space, a class of possible worlds. Subsets of the logical space are propositions, and each interpretation induces a mapping of closed sentences to propositions. The usual logical notions which can be explicated within a possible-worlds ontology (necessary truth, implication, and so on) can, by virtue of the mapping, be transferred from propositions to the sentences which denote them. Thus a sentence A is *necessarily true* (relative to I) just in case I assigns to A the necessarily true proposition. *A implies B* (relative to I) just in case the proposition I assigns to B is true in every world in which the proposition I assigns to A is true. Truth is relative not only to interpretation, but to an interpretation *together* with the state of the world: A is true in W (relative to I) just in case the proposition I assigns to A contains W. A is *true simpliciter* (relative to I) just in case the proposition I assigns to A contains the actual world. But note that which world is actualised is no part of the *interpretation* of a language, the assignment of meanings to terms, and one cannot ascertain the truth value of a sentence from an examination of the interpretation alone. One must check to see which states of affairs obtain in fact.

The notion of *logical* implication, as usually defined in logic textbooks, gives the so-called logical constants a special status: they are the only terms which receive a fixed interpretation throughout, whereas all other terms are allowed to vary in their interpretation. Implication for sentences, as defined here, is relative to a fixed interpretation of *all* terms, not just the logical constants. Thus while all the usual first-order implications will hold in each intensional interpretation, the converse does not hold. A may imply B under interpretation I, but fail to do so under interpretation I', which differs in its assignment of meanings to some terms. Logical implication and interpretation-relative implication thus stand at two different extremes, and we could consider various intermediate notions. Let $\mathscr{I}$ be a class of interpretations of a language. A $\mathscr{I}$-implies B just in case A implies B relative to all interpretations in $\mathscr{I}$. Implication relative to one particular interpretation is the special case in which $\mathscr{I}$ is a singleton. Logical implication is the special case in which $\mathscr{I}$ is the set of all interpretations. There are various intermediate notions. Consider a partition of the vocabulary $\{V_1, V_2\}$, and suppose that I is an interpretation of V_1 over some logical space ω. Let $\mathscr{I}^I$ be the set of all extensions of I to the whole vocabulary. (Obviously, each such extension must use the same logical space as does I.) Implication relative to $\mathscr{I}^I$ is the case in which the members of V_1 have fixed meaning, but the members of V_2 do not. One particularly interesting case will be that generated by the partition of the vocabulary into observational and theoretical, together with a fixed interpretation I_O of the observational part. Call this O-implication.

Return to Ramsey's problem as understood within this intensionalist account. How can the observational content of a theory be extracted, given the positivist assumption that the interpretation of the theoretical terms is problematic, that is, they are not understood prior to their participation in a theory (along with the usual correspondence rules, observational theories, and so on). It is natural to say that B (over V_O) is an observational consequence of A just in case A O-implies B. In the appendix we have a proof of the Ramsey result extended to this intensionalist account of interpretation:

PROPOSITION 3. *A^R has all and only the observational consequences of A.*

An anti-realist of the sort who denies the verification-transcendence doctrine will then be justified in taking A^R to represent the factual content of A. The question arises whether Carnap's method can be extended to extract the analytic component of the theory. As it turns out there is a simple argument to show that if A^R is true in a world W under I_O then there is an extension of I_O, I, to the whole vocabulary, under which the Carnap sentence $A^R \supset A$ is true in W. And so, for any given world W, it is possible to ensure that the theoretical terms are interpreted in such a way that the theory, if observationally adequate, is true in W. But it does not follow from this that the Carnap sentence is appropriately *analytic* under that extended interpretation. For A^C to be analytic (relative to I) it must be true in *all* worlds of the logical space under consideration. We could define weak and strong admissibility, as in the extensional case, but on the intensionalist account there seems to be no good reason to consider weak admissibility. Thus an interpretation is said to be A-admissible if it is *an extension of the intended interpretation of* V_O and A^C is analytic relative to that extension.

Although (as will be shown) A-admissible interpretations do not always exist, they do exist in one large class of cases particularly attractive to positivists: namely, first-order theories. Indeed some positivists would be inclined to impose the condition of being first-order on a theory for it to be acceptable in the first place.

PROPOSITION 4. *If A is a first-order theory then there is an A-admissible interpretation.*

Thus Carnap's method can be extended at least to first-order theories. To say that A^C is to be regarded as analytic is to say that the only legitimate interpretations are those which are A-admissible. Under what conditions are two theories A and B meaning-compatible? Obviously, when there exists at least one interpretation which is both A-admissible and B-admissible. We have the following (intensionalist) analogue of Proposition 2:

PROPOSITION 5. *If A and B are first-order theories then A and B fail to be meaning-compatible just in case there is a world W in which A and B are each observationally adequate, but in which A & B is not observationally adequate.*

It follows immediately from the theorem that observational rivals must be meaning compatible. If *A* and *B* deliver incompatible observational sentences then there can be no world in which both are observationally adequate. When it is remembered that included in *A* and *B* are all the low-level observational laws as well as correspondence rules necessary to generate specific observational predictions it is clear that meaning compatibility is rather extensive. In particular, it will be entirely possible to have rival theories of electrons. Meaning incompatibility arises not so much from *massive* disagreement at the theoretical level, as from rather subtle differences: namely, those which permit piecemeal observational adequacy together with global observational inadequacy.

So far it has been shown that the doctrine of partial interpretation permits a very large degree of meaning compatibility. It remains to show that with or without meaning compatibility the kind of anti-realism espoused by Carnap is rather hospitable to commensurability.

In order to give a reasonably rigorous demonstration of commensurability it is necessary to have to hand a reasonably rigorous account of the comparability of theories for truthlikeness. Although there are a number of theories in existence only one of these has been developed explicitly for higher-order frameworks. In Oddie (1986) it is shown how to define truthlikeness for higher-order systems by means of a theory of higher-order normal forms (called permutative normal forms). Constituents are (possibly higher-order) sentences which normalise the informative content of theories which are maximally informative (relative to some parameters, like quantificational complexity), and all sentences can be shown to be equivalent to disjunctions of constituents. (This result holds whatever kind of equivalence, logical or interpretation relative, or something between, is considered.) Thus every sentence is equivalent to one which gives its information in a highly standardised form. Furthermore, because the syntactic structure of a constituent reflects the structure of the worlds in which it is true, the distance (or likeness) between the structure of worlds can be gleaned from a measure of distance (or likeness) between constituents. A theory is the more truthlike the closer are the constituents in its normal form to the true constituent. However, it is also shown in Oddie (1986) that constituents are suitable for defining distance between structures only if the language in which they are couched is *suitably interpreted.* For a language to be suitably interpreted the primitive predicates must stand for certain special properties: namely, those which generate the logical space, the primary properties.[16] Each logical space is a set of distributions of extensions through some primary or basic properties, and it is only in virtue of this that the space itself has any determinate structure which enables judgements of likeness and distance to be made. And only if the primitive predicates stand for these primary properties (and relations) will the syntactic structure of the constituent reflect the common structure of the worlds in which it is true.

What seems to lie behind the anti-realism of Carnap (and perhaps Putnam and Dummett as well) is the assumption that the space of possibilities at issue in any 'legitimate' inquiry is generated solely by observational properties and relations. That is to say, if two states of affairs are identical as far as the extensions (or values) of observables go, then they are identical *simpliciter.* Put slightly differently, everything supervenes on a base of observational properties. There can be no change in the world without some change in the distribution of observational properties. Given this assumption it is entirely natural to demand that legitimate interpretations be *extensions* of interpretations of the observational vocabulary. Suppose I_O interprets V_O over a space of possibilities ω_O generated entirely by observational properties and relations. Then, since the meanings of theoretical terms are not given independently of the theory, the full factual content of a theory is given by its Ramsey sentence. Consequently, to get at the whole factual content of a theory it is not necessary to countenance logical spaces any more complicated or richer than ω_O itself. If V_T needs to be interpreted at all it may as well be interpreted over the space ω_O. (Note that this way of interpreting Carnapian positivism or anti-realism is available only once we have distinguished between worlds and interpretations in the right way.) Furthermore, it is also entirely natural to demand that the primitive predicates stand for the very observational properties which generate the space ω_O. That is to say I_O is a suitable interpretation of V_O.

It follows that for the purposes of comparing A and B for truthlikeness it is not necessary at all to ascend to the theoretical vocabulary and interpretations of it. A^R and B^R are couched in the observational vocabulary which is, by assumption, suitably interpreted. Moreover, they give the full factual content respectively of A and B. Given that we have an adequate theory of truthlikeness for higher-order sentences it follows that A^R and B^R are comparable for truthlikeness, that is to say, commensurable. This result holds even if A and B are *not* meaning compatible. The existence or non-existence of admissible interpretations of the theoretical aspect of the theories is irrelevant to their comparability for truthlikeness. In order to compare theories for truthlikeness they have to be couched in suitably interpreted languages, and since the only legitimate space (for an anti-realist of the Carnapian kind) is generated by observables, the interpretation of theoretical terms plays no part in such comparisons. From the point of view of commensurability, V_T is redundant, and all sentences formulated over V_O are (relative to I_O) comparable for truthlikeness, whether they are first-order or (like A^R and B^R) higher-order.

In fact it should come as no surprise that Carnap's anti-realism should be hospitable to such widespread commensurability. The whole point of cutting theories down to observational size is to eliminate comparability problems that arise out of the excess metaphysical baggage theories may seem to carry with them.

The real shortcomings of Carnap's anti-realism have nothing to do with alleged meaning variance or alleged incommensurability. Rather, they surface only if one concedes the legitimacy of predicates which are both primitive and higher-order. Given such predicates it may be impossible to extend I_O to an interpretation which is A-admissible. Hard-nosed empiricists may well not be moved by this. The idea that theories should incorporate primitive higher-order terms is not one which they would normally endorse. However, there are one or two hard-nosed empiricists who have recently been forced to admit just this. David Armstrong (along with Fred Dretske and Michael Tooley) has recently argued that laws of nature must be analysed in terms of natural necessitation, and that this is a higher-order relation between properties.[17] Armstrong himself also argues that properties cannot be analysed away as purely extensional entities (a position he calls class nominalism), so that the items which are related by necessitation are not classes. Moreover, natural necessitation implies constant conjunction but cannot be analysed away by means of constant conjunction. It is not far from here to the thesis that natural necessitation is one of the relations which generate the space of possibilities of physics, or science in general.

Let us assume, for the moment, this Armstrongian theory of natural necessitation. Consider the following simple theory:

A: $N(P, T) \& N(T, Q)$.

Let us suppose that P, Q and N all belong to that part of the vocabulary which is antecedently understood (and that N stands for the

Armstrongian relation of natural necessitation) while T is not understood independently of the theory A. On certain interpretations of P and Q it may well be the case that there are worlds W and V in which the Ramsey sentence of A, A^R (that is $(\exists t)(N(P, t) \& N(t, Q)))$ is true even though there is no property which, when assigned to t, satisfies the open sentence $(N(P, t) \& N(t, Q))$ in *both* W and V. Because of the rejection of class nominalism, whether or not natural necessitation holds between two properties, depends on more than the extensions of the properties. As such there will be no extension of I_O which ensures the truth of $A^R \supset A$ in both W and V. Consequently, there is no interpretation which renders the Carnap sentence analytic: there are no A-admissible interpretations. This would hardly upset a confirmed anti-realist of the Carnap-Putnam variety. Rather, it would be taken as proof that the relation of natural necessitation should not be countenanced.

There is, then, no hard and fast connection between anti-realism and incommensurability. The anti-realist of the Carnap-Putnam variety can espouse quite a robust version of commensurability. Moreover, this result is quite independent of the question of the meaning of theoretical terms. As it turns out, most rival theories are meaning compatible. But even if they are not it is quite in order, for thc anti-realist, to compare Ramseyfied theories alone, since these contain all the factual content on the underlying positivist assumptions. Moreover, there is available an adequate theory of truthlikeness for higher-order theories which ensures commensurability in these cases.

4. REALISM AND INCOMMENSURABILITY

Realism, by virtue of its commitment to the possibility of progress, is also committed to a fairly robust version of commensurability. What will be shown here is that theories which it is extremely tempting to regard as commensurable are in fact not so, and that this incommensurability which is rather widespread, is demanded by realism itself.

According to realists, the aim of an inquiry is the truth about (some aspect of) the structure of the world.[18] Suppose that we are interested solely in the possession or otherwise of a property **P** and that in fact all objects in the universe possess this property, save one. The single exception is **X**. Now consider the following two theories:

A: $(\forall x)P(x)$

B: $(\forall x) \sim P(x)$

Let us suppose that P is interpreted over a framework in which **P** is the sole generating trait, and that P stands for that very property. Intuitively, A is closer to the truth than is B. If anything is progress towards the truth then the step from B to A in these circumstances is progress towards the truth.

The actual world, $\mathbf{W}_0$ can be depicted as in the first diagram, with shading representing possession of **P**, and **X** to the far left of the diagram,

```
        X
        ↓
W0      ○ ● ● ● ● ... (etc.) ...
```

A and B, under the interpretation to hand, are true in just one state of affairs each, respectively $\mathbf{W}_1$ and $\mathbf{W}_2$.

```
        X
        ↓
W1      ● ● ● ● ● ... (etc.) ...
        X
        ↓
W2      ○ ○ ○ ○ ○ ... (etc.) ...
```

The reason A is closer to the truth than B is that the structure picked out by A (that is, $\mathbf{W}_1$) is closer to the actual structure than is that picked out by B (that is $\mathbf{W}_2$).

There are conceptual schemes other than this one. Consider what seems to be a highly contrived property, **Q**. Firstly, let **C** be the following condition: for any object X, X has the property **P** just in case **X** lacks **P**. Note that **C** holds in $\mathbf{W}_0$ but fails in $\mathbf{W}_1$ and in $\mathbf{W}_2$. Now let **Q** be the property which an object X has just in case the following holds: X has **P** if and only if **C** holds. **Q** has an admittedly strange characterisation, but there seems to be no good logical ground for ruling it out. In particular, if we had started with **Q** we could have characterised **P** in an exactly analogous fashion. For let **D** be the following condition: for every X, X has **Q** just in case **X** lacks **Q**. Then

P holds of X just in case the following condition holds: X has **Q** if and only if **D** holds. Consider various distributions of **Q** through the same domain of objects as that considered above.

$\mathbf{U}_0$ ○ ⊕ ⊕ ⊕ ⊕ ... (etc.) ...

$\mathbf{U}_1$ ○ ○ ○ ○ ○ ... (etc.) ...

$\mathbf{U}_2$ ⊕ ⊕ ⊕ ⊕ ⊕ ... (etc.) ...

Now it may well be tempting to think that $\mathbf{U}_0$ is the very same state of affairs as $\mathbf{W}_0$, $\mathbf{U}_1$ is the same as $\mathbf{W}_1$, and $\mathbf{U}_2$ is the same as $\mathbf{W}_2$. And in general it may be tempting to think that each distribution of **P** through the domain is identical to some corresponding distribution of **Q** through the domain, so that these apparently distinct logical spaces really contain the *same* states of affairs. Let us trace the consequences of this view, which we may call the *identity thesis*.

Consider the following two theories:

A': $(\forall x) \sim Q(x)$

B': $(\forall x) Q(x)$.

Let Q be interpreted over the **Q**-space and let Q stand for **Q**. Then A' is true in just one state of affairs, namely $\mathbf{U}_1$, and B' is true in just $\mathbf{W}_2$. Moreover, according to our story, $\mathbf{U}_0$ must be actualised. If the identity thesis is correct then A and A' (respectively, B and B') are true in exactly the same states of affairs. They have the same truth conditions, or express the same proposition. But from the point of view of the **Q**-space it now appears that A' is further from the truth than is B', and so if the respective identities hold, our judgements of truthlikeness have been reversed. A' gets nearly everything wrong while B' gets nearly everything right. The change from **P** to **Q** seems to have completely reversed the relationships of closeness to truth.

This argument is one of a kind which David Miller has urged in a series of papers aimed to undermine accounts of truthlikeness which take seriously the notion of *likeness* to truth.[19] Moreover, it is now clear that this type of argument is completely general and that, if sound, would destroy all the intuitive judgements of truthlikeness which turn on the idea of the closeness of fit of a theory to facts.[20] In fact Miller's argument admits of a generalisation which is much neater than the above and is easier to grasp than almost any of its particular instances.[21]

A way of conceptualising the world generates a set of possibilities, and these possibilities can usually be thought of as some kind of mathematical distributions. For example, in physics a state space is often generated by a set of real valued magnitudes. A possible state at a moment is an assignment of values to these magnitudes. This in turn is usually identified with an N-tuple of real numbers. A full-bodied set of possibilities is generated by the set of functions from the set of time points to such instantaneous states. However, we can just as easily think of the very same space of possibilities as an assignment of such functions to a global 'magnitude', ϕ. Thus functions from the set of time points into $\mathbb{R}^N$ are possible values, or extensions, of the global magnitude ϕ, and each world is represented by an ordered couple $\langle \phi, f \rangle$ where f is one of the possible values of ϕ. Let $\mathscr{F}$ be the collection of all such functions. Since $\mathscr{F}$ has a mathematical structure it is possible to construct various kinds of measures on it which may be useful in characterising closeness to the truth. Suppose that by virtue of the structure of $\mathscr{F}$, f_1 is intuitively closer to f than is f_2, and that f is the extension, or actual value, of the global magnitude ϕ. The whole truth is thus the theory that $\phi = f$. The theory that $\phi = f_1$ is intuitively closer to the truth than the theory that $\phi = f_2$. Now consider the following one-to-one mapping on $\mathscr{F}$:

$$t(g) = \begin{cases} f_1 & \text{if } \; g \text{ is } f_2 \\ f_2 & \text{if } \; g \text{ is } f_1 \\ g & \text{otherwise.} \end{cases}$$

We can define a distinct global magnitude ψ. The extension of ψ is the t of the extension of ϕ. The set of possibilities generated by ψ also uses the class $\mathscr{F}$, and each possibility is an ordered couple $\langle \psi, f \rangle$, f a member of $\mathscr{F}$. According to the identity thesis $\langle \phi, f \rangle$ is identical to $\langle \psi, f \rangle$, $\langle \phi, f_1 \rangle$ is identical to $\langle \psi, f_2 \rangle$ and $\langle \phi, f_2 \rangle$ is identical to $\langle \psi, f_1 \rangle$. Thus the truth is that $\psi = f$, the theory that $\phi = f_1$ is equivalent to the theory that $\psi = f_2$, the theory that $\phi = f_2$ is equivalent to the theory that $\psi = f_1$. If the identity thesis holds then not only are judgements of truthlikeness reversed. So too are all judgements which depend in any way at all on the structure of the set of functions $\mathscr{F}$. As is shown in detail in Oddie (1986), judgements of confirmation and disconfirmation, change, structure, and similarity in general, all dissolve in the face of this kind of generalised Miller argument.[22]

Although we have been assuming that $\mathscr{F}$ underlies the kind of possibility space considered in physics nothing in the argument hangs on this. $\mathscr{F}$ can be any class of structures which may feature as the values or possible extensions of the global state, *the actual world*. Let us call transformations like *t*, and that used in the **P-Q** example, *Miller-transforms.*

It is clear that the realist does not have all that many options in the face of Miller's argument. The fundamental choice is that of either accepting or rejecting the identity thesis: that logical spaces related by Miller-transforms are identical, and that theories interpreted over one space are intertranslatable with theories interpreted over a Miller-transform of that. Acceptance of the identity thesis virtually annihilates the possibility of progress in one blow. Since Miller-type arguments can be used to destory *all* the intuitive judgements on the relative closeness of theories to some specified state of affairs, if we grant the identity thesis then there would be no adequacy conditions left for a rigorous account of truthlikeness. The only kind of account of truthlikeness compatible with the identity thesis is that according to which no false theory is closer to the truth than any other. As such the identity thesis implies a very strong form of incommensurability: false theories simply cannot be graded for truthlikeness.

The alternative is to reject the identity thesis and to deny the intertranslatability of theories based on different conceptual schemes. On this view A is closer to the truth than B, and A' is further from the truth than B'. There is no contradiction here, because A and A' (and B and B') stand for entirely different propositions. They say entirely different things. The set of worlds generated by the property **P** is distinct from the set of worlds generated by the theory **Q**, and propositions over the first space have a completely different assertive force to any over the second space. Theories interpreted over the **P**-space are commensurable with each other, as are theories interpreted over the **Q**-space. But theories over the **P**-space are not commensurable with those over the **Q**-space.

The result of rejecting the identity thesis is a kind of *restricted* incommensurability. There is commensurability *within* conceptual frameworks, but not *across* conceptual frameworks. Theories based on different conceptual frameworks have entirely different assertive forces. This conclusion may be unpalatable to many realists, but the only viable alternative seems to be the *unrestricted* incommensurability which rides in the wake of the identity thesis.

There are two other responses open to realists which may seem viable. One is to claim that a theory of truthlikeness can be constructed which does not utilise any of the intuitive judgements which are destroyed upon acceptance of the identity thesis. This seems highly unlikely given the degeneration of the only programme for defining truthlikeness which rivals the likeness programme: namely the probability-content-programme.[23]

The other is to claim that while the **P**-space and the **Q**-space do indeed contain exactly the same states of affairs nevertheless judgements of truthlikeness (change, structure and so on) generated by only one of these spaces can be correct. The world itself selects one of these spaces as privileged. There is, in nature itself, a preferred system of concepts. The resulting doctrine, call it *super-realism*, is not entirely unattractive, but it hardly needs stressing that it involves some rather severe problems. Two can be briefly sketched. Firstly, nature's preference for **P** over **Q**, say, would have to be either a matter of necessity or not. In either case there seems to be a real problem stating what, in nature, would ground the preference. Of course, the position can simply be insisted on in these or slightly different words, but that would not be philosophically illuminating. Secondly, if the preference is a matter of necessity then our access to the privileged or preferred properties or concepts would have to be *a priori*. If the preference is a contingent matter then it would have to be *a posteriori*. Again, there are obviously severe problems showing how it could be either. This does not by any means disprove super-realism, but the onus is on the super-realist to supply an account.

Eschewing super-realism, then, the choice is between restricted and unrestricted incommensurability. If, as I have argued, the doctrine of the possibility of progress is central to any reasonable version of realism, it follows that realists must choose restricted incommensurability, as the price for denying the identity thesis and affirming the coherence of the concept of truthlikeness.

It may be charged that the rejection of the identity thesis is *ad hoc*, an expedient for saving realism in the face of a powerful objection to the doctrine of the possibility of progress. In fact, the rejection of the identity thesis can be motivated from a consideration of the core doctrine of realism, the truth doctrine. If the aim of an inquiry is the truth about (some aspect of) the structure of the world, then it follows that for the aim to be coherent the world must have a structure. If we can show that this alone entails the rejection of the identity thesis then the charge of *ad hocery* fails.

Although it may happen that no other possible world shares the structure of the actual world, it is not necessarily so. Structure is something which different possible state of affairs may have in common. If $\mathbf{W}_1$ is the actual world then, as it happens, it is the only world with that particular structure. All worlds with the same structure as $\mathbf{W}_1$ are identical to $\mathbf{W}_1$. On the other hand, $\mathbf{W}_0$ shares its structure with a number of distinct possible worlds. For example, consider any other distribution of **P** which assigns the property to all bar one individual. $\mathbf{W}_3$ is an example — it endows everything except the individual **Y** (distinct from **X**) with **P**.

Y
↓
$\mathbf{W}_3$ ● ○ ● ● ● ... (etc.) ...

$\mathbf{W}_0$ and $\mathbf{W}_3$ do not differ over the structure of their distributions of **P**, only over which individuals have it. This concept of structural identity of worlds can be given a rigorous explication [as for example Carnap has done in his (1971)].[24] Now consider distributions of **Q** through the domain. If the identity thesis were correct then $\mathbf{U}_3$ would be the very same state of affairs as $\mathbf{W}_3$:

Y
$\mathbf{U}_3$ ○ ⊕ ○ ○ ○ ... (etc.) ...

It is apparent that $\mathbf{U}_0$ and $\mathbf{U}_3$ are not structurally identical. They differ substantially in their structure. Now the following four judgements are logically incompatible:

$$\mathbf{W}_0 = \mathbf{U}_0, \text{ the structure of } \mathbf{W}_0 = \text{the structure of } \mathbf{W}_3,$$
$$\mathbf{W}_3 = \mathbf{U}_3, \text{ the structure of } \mathbf{U}_0 \neq \text{the structure of } \mathbf{U}_3.$$

If the world does have a structure then the concept of structure is not spurious. On the other hand if the identity thesis is correct the notion of structure is spurious, for it leads to the above contradiction. Hence we must either give up the idea that the world has a structure, or give up the identity thesis. Realists who eschew super-realism must give up the identity thesis. If the two different spaces do indeed contain two completely different sets of structures, then it is not at all surprising that theories based on these different spaces are not commensurable. The demand for (restricted) incommensurability can thus be seen to arise quite naturally from the core doctrine of realism.[25]

5. CONCLUSION

The realist is, then, committed to a restricted version of the thesis of incommensurability. Theories interpreted over different conceptual frameworks, or logical spaces, may appear to be rivals, but because they cannot be compared for truthlikeness they are nevertheless incommensurable. Anti-realists, like Carnap, who are prepared to accept the doctrines of truth and the possibility of progress will also be saddled with this restricted incommensurability. But the realist cannot avoid this much incommensurability, and the anti-realist is not forced to accept more.

While this alone would be sufficient to undermine the dogma, the situation may be worse for the realist, and better for the anti-realist, than this suggests. The anti-realist does not have to dispense with the theoretical vocabulary, but he does show that it is not essential for capturing the full factual content of a theory. Once theories are pared down to a common observational vocabulary there is nothing to prevent them being interpreted over a common logical space the generating properties of which are all observational. On the other hand, realists will usually want to claim that a theory's factual content transcends its observational content, and that the theoretical terms do not receive their interpretation solely by virtue of the role they play in the theory. (For example, we seem to be able to connect the term 'cause' to the concept of causation, and if the history of the repeated failures to reduce this to observables is anything to go by, the causal relation is not simply observable.) As such it is not at all unlikely that two different theories, apparently in the same domain, will employ different theoretical concepts, and so be interpreted over different logical spaces. Such theories would then be incommensurable. Anti-realists who insist on the partial interpretation of theoretical terms may thus well be in a far better position to enforce commensuration than realists who insist that such terms can be interpreted in a theory-independent way.

APPENDIX

PROPOSITION 1. *A and B fail to be weakly meaning compatible if and only if they are logically incompatible and both are observationally uncreative.*

Proof. Suppose there is no interpretation which is both weakly A-admissible and weakly B-admissible. That is, there is no interpretation in which both A^C and B^C are true. Hence $A^C \models \sim B^C$. So $A^C \models \sim(B^R \supset B)$, hence $A^C \models B^R \& \sim B$. Because $A \models A^C$, it follows that $A \models \sim B$, so that A and B are logically incompatible. Now suppose that $B \models S_O$. Then $B^R \models S_O$. But by the above $A^C \models B^R$, so that $A^C \models S_O$. But A^C implies no non-trivial observation sentence. Hence, S_O is trivial, that is $\models S_O$. Similarly we can show that if $A \models S_O$ then S_O is trivial, by an analogous argument.

Now suppose that A and B are logically incompatible and observationally uncreative. For the sake of a *reductio* assume that I is both weakly A-admissible and weakly B-admissible. Then $A^R \supset A$ is true in I as is $B^R \supset B$. Since A (B) is observationally uncreative, all of its observational consequences are trivial, and hence true in all interpretations. Thus A^R (B^R) is true in all interpretations. It follows that A and B must also be true in I, contrary to the hypothesis that A and B are logically incompatible.

PROPOSITION 2. *A and B fail to be strongly meaning compatible if and only if A and B are observationally adequate while A &B is not observationally adequate.*

Proof. Suppose there is no interpretation which is both strongly A-admissible and strongly B-admissible. So there is no extension of I_O in which both A^C and B^C are true. Assume, for the sake of a *reductio*, that A is not observationally adequate. Hence A^R is false in I_O. A^R is false in every extension of I_O, and so $A^R \supset A$ is true in every extension of I_O. That is to say, all extensions of I_O are strongly A-admissible. Hence no extensions are strongly B-admissible (since none are both). In all extensions of I_O B^C is false, and so B^R is true while B is false. It follows that B^R is true in I_O, but then (by the argument in the text of Section 3) there is at least one extension of I_O in which B is true. Consequently at least one such extension is strongly B-admissible. Contradiction. By an analogous argument we can show that B is observationally adequate. Now for the sake of a *reductio* suppose that

$A \,\&\, B$ is observationally adequate. Then $(A \,\&\, B)^R$ is true in I_O, and so there is at least one extension of I_O in which $A \,\&\, B$ is true. In this extension both $A^R \supset A$ and $B^R \supset B$ are true, and so it is both strongly A-admissible and strongly B-admissible.

Now suppose that A and B are both observationally adequate and $A \,\&\, B$ is not. Hence A^R and B^R are both true in I_O and $(A \,\&\, B)^R$ is false. Suppose I is both strongly A-admissible and strongly B-admissible. Then $A^R \supset A$ is true in I as is $B^R \supset B$. Consequently A and B are also true in I, as is $A \,\&\, B$. $A \,\&\, B \vDash (A \,\&\, B)^R$, so that the latter is true in I. Contradiction. Hence no interpretation is both strongly A-admissible and strongly B-admissible.

In order to prove propositions 3, 4 and 5 we need an intensionalist account of interpretation for higher-order languages. Here we will concentrate on a class of languages which are relatively simple, easily manageable, extensions of usual first-order ideography, but which are nevertheless far more powerful in their expressive power than any first-order language. For example, it is possible to express in them facts about Armstrong's relation of natural necessitation.[26]

Each logical space, or set of worlds, ω, together with the domain of lowest level entities (individuals) ι, and the set of truth values, o, generates a hierarchy of *types* of entities. The type hierarchy is characterised recursively. ω, ι, and o are all types. Where $\xi_1, \ldots, \xi_n$ are any types, a $\xi_1, \ldots, \xi_n$-string is an n-tuple of objects the ith member of which is an object of type ξ_i. Where $\eta, \xi_1, \ldots, \xi_n$ are any types $(\eta\xi_1 \ldots \xi_n)$ is the type of all functions taking $\xi_1, \ldots, \xi_n$-strings into objects of type η.

This gives us a convenient typology of familiar objects. Sets of ξ-objects, for example, are of type $(o\xi)$: sets of individuals, $(o\iota)$; sets of worlds (propositions), $(o\omega)$; sets of propositions, $(o(o\omega))$. Properties of objects of type ξ are of type $((o\xi)\omega)$; properties of individuals, $((o\iota)\omega)$; properties of propositions, $((o(o\omega))\omega)$, and so on. Extensional relations (or linkages) are of type $(o\xi_1 \ldots \xi_n)$: dyadic linkages between individuals, $(o\iota\iota)$; dyadic linkages between individuals and propositions, $(o\iota(o\omega))$; and so on. Relations-in-intension (or just relations) are of type $((o\xi_1 \ldots \xi_n)\omega)$; dyadic relations between individuals, $((o\iota\iota)\omega)$; relations between individuals and propositions, $((o\iota(o\omega))\omega)$; and so on. Properties are special cases of relations ($n = 1$), as are propositions (the degenerate case in which $n = 0$).

The higher-order languages to be constructed will enable all types of objects over a logical space to be discussed, in the same way that first-order languages enable individuals to be discussed. They share the following symbols in common with first-order languages: &, $\vee$, $\sim$, $\exists$, $\forall$, (,), the comma. In addition there is an identity sign $=^{\xi}$ for each type ξ, and an infinite set of variables, x_1^{ξ}, x_2^{ξ}, ... for each type ξ. Different higher-order languages differ in their vocabularies. A vocabulary consists of a collection of symbols, each of which is associated with a particular type — these will be called *constants*, and both variables and constants will be called *terms*. Terms of type $(o\xi_1 \ldots \xi_n)$ for some types $\xi_1, \ldots, \xi_n$ will be called *linkage terms*, and those of type $((o\xi_1 \ldots \xi_n)\omega)$ will be called *relation terms*. Extensional logic does not distinguish between linkages and relations, and there is no way within ordinary extensionalist semantics of mimicking the role played by higher-order relations, even in higher-order (extensional) logic.

The class of atomic formulae is defined recursively: (i) Where $a_1, \ldots, a_n$ are any terms of types $\xi_1, \ldots, \xi_n$ respectively, and b is either of the type $(o\xi_1 \ldots \xi_n)$ or $((o\xi_1 \ldots \xi_n)\omega)$ then $b(a_1, \ldots, a_n)$ is an *atomic formula*. Steps (ii) and (iii) of the recursion are just the same as that for first-order languages, allowing non-atomic formulae to be built up from other by means of the usual truth functional terms and quantification. Here, of course, quantification may be over any type.

An *interpretation* of a vocabulary V is an assignment to each member of V of type ξ an object of type ξ (over some logical space), for each type ξ. We include in this the assignment of the identity linkage of type $(o\xi\xi)$ to $=^{\xi}$ for each ξ. (A *suitable* interpretation is one which pairs off primitive relation constants with the relations which correspond to those that generate the space in question, but this concept plays no role in the following formal proofs.) A *valuation* J is an assignment of objects to all variables of the language, objects of type ξ being assigned to variables of type ξ.

Given an interpretation I each formula of the language can be associated with a proposition, relative to a valuation J. Where a is any term let $O_J(a)$ be the object assigned to a by I under valuation J: if a is a variable $O_J(a)$ is just $J(a)$, if it is a constant $O_J(a)$ is $I(a)$. Furthermore, where a is any term, let $E_J(a, W)$ be the *extension* of $O_J(a)$ in W: if a is of the type $(\xi\omega)$ for some ξ then $E_J(a, W)$ is the value of $O_J(a)$ at W; otherwise, it is just $O_J(a)$ itself. (Intuitively, intensions are objects of type $(\xi\omega)$ for some ξ.)

The recursive steps in the association of propositions with formulae are now rather obvious. (i) If A is atomic, and of the form $b(a_1, \ldots, a_n)$ then the proposition I associates with A, relative to valuation J, is the set of worlds in which the objects assigned to $a_1, \ldots, a_n$ form a string (in that order) which is in the extension of the object assigned to b: $Pr_I(A, J) = \{W: \langle O_J(a_1), \ldots, O_J(a_n)\rangle \in E_J(b, W)\}$. (ii) if A is of the form $B \& C$ then $Pr_I(A, J)$ is the intersection of $Pr_I(B, J)$ and $Pr_I(C, J)$. If A is of the form $\sim B$ then $Pr_I(A, J)$ is the complement of $Pr_J(B, J)$, and so on for the other truth functional connectives. If A is of the form $(\exists x^{\xi})B$ then $Pr_I(A, J) = \{W: (\exists X)(X$ is of type ξ and $W \in Pr_I(B, J[X/x^{\xi}]))\}$ (where $J[X/x^{\xi}]$ differs from J at most in its assignment of X to x). An analogous definition applies to universal quantification.

It can be easily proved that if A has no free variables then the assignment of a proposition to A is independent of the valuation J. In this case we can speak of the proposition assigned to A *simpliciter*, $Prop_I(A)$. All the usual notions applicable to propositions, considered as sets of worlds, can now be applied to sentences with which they are associated. Hence A implies B (relative to I) just in case $Prop_I(A)$ implies (is a subset of) $Prop_I(B)$. A is equivalent to B just in case $Prop_I(A)$ is the very same as $Prop_I(B)$. Note that we will have the usual first-order implications under any interpretation, as well as truth functional laws, and the usual laws of quantification. It would take considerable space to prove all these in a completely rigorous manner, and in what follows such laws, together with various obvious principles of substitution, will be taken for granted. In particular, Lemma 1 can be proved by induction. Let I be an interpretation of all the constants in $A(t_1/T_1, \ldots, t_m/T_m)$ and let I^J be that extension of I which assigns to T_i the same object that J assigns to t_i (for each i).

LEMMA 1. *For any I and J,* $Pr_{I^J}(A, J) = Pr_I(A(t_1/T_1, \ldots, t_m/T_m), J)$.

And this yields as a corollary the first half of the extension of Ramsey's theorem: for every I, A implies$_I$ A^R.

COROLLARY *A logically implies* A^R.

Recall that where I_O is the intended interpretation of the observational vocabulary, and $\mathscr{I}_O$ is the class of all extensions of I_O to the theoretical vocabulary, A O-implies B just in case for very I in $\mathscr{I}_O$ A implies$_I$ B. An S_O is an observational consequence of A just in case S_O contains no constants outside the observational vocabulary, and A O-implies S_O. (Note that A may well O-imply a sentence that it does not logically imply, because the interpretation of the observational vocabulary will undoubtedly forge necessary connections between such terms.)

PROPOSITION 3. *A^R has all and only the observational consequences of A.*

Proof. The corollary above gives us one half of the theorem. If A^R O-implies S_O, then for every I in $\mathscr{I}_O$ $Prop_I(A^R)$ is a subset of $Prop_I(S_O)$. By the corollary, for every I, $Prop_I(A)$ is a subset of $Prop_I(A^R)$. Now suppose that A O-implies S_O, and for the sake of reductio, that A^R does not. Then there is an I in $\mathscr{I}_O$ such that $Prop_I(A^R)$ is not a subset of $Prop_I(S_O)$. Because neither A^R nor S_O contain terms which are not observational, and I makes the same assignments to these as does I_O, it follows that $Prop_{I_O}(A^R)$ is not a subset of $Prop_{I_O}(S_O)$. Hence there is a world in the former not in the latter, say W. By the recursive definition of Pr it follows that there must be a valuation J such that W is in $Pr_{I_O}(A(t_1/T_1, \ldots, t_m/T_m), J)$.

By Lemma 1, $Pr_{I'_O}(A, J) = Pr_{I_O}(A(t_1/T_1, \ldots, t_m/T_m, J)$. By the fact that A contains no free variables it follows that $Prop_{I'_O}(A) = Pr_{I'_O}(A, J)$, so that W is a member of $Prop_{I'_O}(A)$. Because S_O contains only observational constants, and W is not in $Prop_{I_O}(S_O)$, it follows that W is not in $Prop_I(S_O)$ for any extension I of I_O. In particular W is not in $Prop_{I'_O}(S_O)$. But then A does not imply$_{I'_O}$ S_O, and so does not O-imply S_O. Contradiction.

We need another lemma in order to prove Proposition 4 and 5. A is said to be *first-order* just in case all the terms that occur in A are of the type of an individual ι or of a relation between individuals (including the special cases of properties and propositions).

LEMMA 2. *If for any constant or free variable a in A, $E^I_J(a, W) = E^{I'}_{J'}(a, W)$, then if A is first-order, W is in $Pr_I(A, J)$ just in case W is in $Pr_{I'}(A, J')$.*

Proof. Let A be of the form $b(a_1, \ldots, a_n)$. Since A is first-order the a_is must be of the type of an individual, and so the condition ensures that $O^{I'}_{J'}(a_i) = O^{I}_{J}(a_i)$. b is of the type of an n-place relation between individuals, and the condition ensures that $E^{I}_{J}(b, W) = E^{I'}_{J'}(b, W)$. Jointly these two facts yield the desired result. The steps for the truth functions are obvious. Let A be $(\exists x)B$. W is in $Pr_I(A, J)$ just in case there is an X such that W is in $Pr_I(B, J(X/x))$. $J(X/x)$ assigns to x the same object as does $J'(X/x)$, and so for all terms free in B, $E^{I}_{J(X/x)}(a, W) = E^{I'}_{J'(X/x)}(a, W)$. By inductive hypothesis the desired result follows. Similar considerations apply to universal quantification.

PROPOSITION 4. *If A is first-order then there is an A-admissible interpretation.*

Proof. We must find an extension of I_O in which $A^R \supset A$ is necessarily true: for all W, W is a member of $Prop_I(A^R \supset A)$. I must assign a relation to each term T_i, X_i. Since relations are functions from worlds to linkages (sets, truth values) it follows that X_i is specified once its extension in each possible world is specified. (i) Suppose W is in $Prop_{I_O}(A^R)$. Hence there is a J such that W is a member of $Pr_{I_O}(A(t_1/T_1, \ldots, t_m/T_m), J)$, and so, by Lemma 1, W is a member of $Pr_{I^J_O}(A, J)$. Let I assign to T_i an intension X_i whose extension in W is the same as that of $I^J_O(T_i)$ $(= J(t_i))$. By Lemma 2 W is in $Pr_I(A, J)$. Since A contains no free variables W is in $Prop_I(A)$. (ii) Suppose W is not in $Prop_{I_O}(A^R)$. Then there is no extension I of I_O in which W is a member of $Prop_I(A^R)$. Hence it does not matter what the extension of X_i is in W, since whatever it is W must be a member of $Prop_I(A^R \supset A)$. Provided each X_i is constructed in accordance with steps (i) and (ii) then if I assigns X_i to T_i, W is in $Prop_I(A^R \supset A)$, for every world W.

PROPOSITION 5. *If A and B are first-order then A and B fail to be meaning compatible just in case there is a world W in which A and B are each individually observationally adequate, but in which $A \& B$ is not observationally adequate.*

Proof. Suppose there is a world W such that W is in $Prop_{I_O}(A^R)$ and in $Prop_{I_O}(B^R)$ but not in $Prop_{I_O}((A \& B)^R)$. Suppose that I is both A-admissible and B-admissible. Then both $A^R \supset A$ and $B^R \supset B$ are necessarily true relative to I, so that W is in both $Prop_I(A^R \supset A)$ and $Prop_I(B^R \supset B)$ and so in both $Prop_I(A)$ and $Prop_I(B)$, and so in $Prop_I(A \& B)$; by the corollary, W is in $Prop_I((A \& B)^R)$: contradiction.

Now suppose that A and B are not meaning compatible. Call a world which is a member of $Prop_{I_O}(A^R)$ (respectively: $Prop_{I_O}(B^R)$) an A^R-world (respectively: a B^R-world). Note that the construction of an A-admissible interpretation (respectively: a B-admissible interpretation) does not impose any constraints on the extensions of the terms in worlds which are not A^R-worlds (respectively: not B^R-worlds). Hence if no world is both an A^R-world and a B^R-world there is no impediment to constructing an interpretation which is both A-admissible and B-admissible. And so there must be some worlds in which both A^R and B^R are true, that is, in which both A and B are individually observationally adequate. Let two interpretations I and I', such that for each i the extension of $I(T_i)$ in W is the same as the extension of $I'(T_i)$ in W, be called *W-equivalent*. By Lemma 2, if I and I' are W-equivalent extensions of I_O then for any first-order formula C, W is in $Prop_I(C)$ just in case W is in $Prop_{I'}(C)$.

Now, for the sake of *reductio* suppose that for any W, if W is in $Prop_{I_O}(A^R)$ and in $Prop_{I_O}(B^R)$, then W is in $Prop_{I_O}((A \& B)^R)$. We can then construct an interpretation I which satisfies the following conditions: (i) for each $(A \& B)^R$-world W, I is W-equivalent to I_O^J for some J such that W is in $Pr_I((A \& B)(t_1/T_1, \ldots, t_m/T_m), J)$; (ii) for each A^R-world W which is not a B^R-world, I is W-equivalent to I_O^J for some J such that W is in $Pr_{I_O}(A(t_1/T_1, \ldots t_m/T_m), J)$; (iii) for each B^R-world W which is not an A^R-world, I is W-equivalent to I_O^J for some J such that W is in $Pr_{I_O}(B(t_1/T_1, \ldots, t_m/T_m), J)$.

It follows that if W is an $(A \& B)^R$-world, W is in $Prop_I(A \& B)$, hence W is in $Prop_I(A)$ and in $Prop_I(B)$, and so is in $Prop_I(A^R \supset A)$ and in $Prop_I(B^R \supset B)$. If W is an A^R-world but not a B^R-world, W is in $Prop_I(A)$ and so is in $Prop_I(A^R \supset A)$; moreover, W is clearly in $Prop_I(B^R \supset B)$. If W is a B^R-world, but not an A^R-world, W is in $Prop_I(B)$ and so is in $Prop_I(B^R \supset B)$; moreover, W is clearly in $Prop_I(A^R \supset A)$. Consequently, whatever category W falls into, it is in both $Prop(A^R \supset A)$ and $Prop(B^R \supset B)$ — I is both A-admissible and B-admissible, contrary to the initial assumption that A and B are not meaning-compatible. Thus on the assumption that A and B are not meaning-compatible it follows that there is a world W which is an A^R-world and a B^R-world, but not an $(A \& B)^R$-world.[27]

Massey University

NOTES

[1] That this doctrine of verification transcendence is at the heart of the realist/anti-realist debate receives strong support in Smart (1986).
[2] For a more detailed account of the necessity for realists to give an explication of truthlikeness, see Oddie (1986), pp. 1—4, and 10—20.
[3] Devitt (1979), p. 29.
[4] Musgrave (1979), p. 336.
[5] English (1978), p. 58.
[6] Popper (1963), p. 234.
[7] As is now well known, Popper's simple idea turned out to be inadequate, in the sense that it fails to deliver intuitively obvious judgements on false theories. Rather ironically it has the consequence that *no* two false theories are comparable for truthlikeness: unmitigated incommensurability. See Tichý (1974), and for a longer discussion, Oddie (1986), chapter 2.
[8] See Carnap (1966), chapter 28. For an extremely clear exposition together with some interesting additional results, see Winnie (1975).
[9] That Carnap's proposal does entail meaning-variance is argued in English (1978), as well as Musgrave (1979), but the most detailed technical paper on these lines is Williams (1973). Williams proves what appears to be the exact opposite of my claim here, and there is nothing wrong with that proof. For further clarification, see footnote 15.
[10] For an extension of the Ramsey-Carnap method to infinite postulate systems, see Tichý (1970).
[11] See the excellent exposition in Winnie (1975).
[12] Musgrave (1979), p. 342 (quoting Hempel).
[13] Carnap (1971), p. 56: note that Carnap uses the term 'model' for what I here call a 'world'.
[14] For further arguments for the distinction between worlds and interpretations, see Oddie (1986), pp. 34—8, 60—6, 152—5.
[15] It is by virtue of this definition of meaning compatibility that results which might appear to conflict with those in Williams (1973) are derived. There is no formal contradiction however — it is just that Williams would probably consider this definition too weak. Certainly his conditions for *intertranslatability* (p. 360) are much more stringent. However, my definition of meaning *compatibility* is not meant to capture *meaning identity*. And once we take the *partiality* in Carnap's account of partial interpretation seriously, then this seems entirely natural. The question of strict intertranslatability arises only with respect to two *complete* specifications of meaning, that is, two determinate interpretations of the relevant vocabulary.
[16] See Oddie (1986), pp. 34—8, 60—5.
[17] See Armstrong (1978) and (1983).
[18] A version of the ensuing argument was presented in Oddie (1981). Note that in this version the positional properties have been eliminated not only for simplicity, but also to make the example virtually indistinguishable from Miller's example in his (1974).
[19] See Miller (1974), (1975) and (1978).
[20] See Oddie (1986), chapter 6.

[21] This generalisation is suggested by an argument in Tichý (1978a), p. 194.
[22] Oddie (1986), chapter 6.
[23] See Oddie (1986), chapters 2 and 3.
[24] See Carnap (1971), p. 118.
[25] The version of the structure argument outlined in Oddie (1981) was criticised in Pearce (1983) and Urbach (1983). As is shown in Oddie (1986), these criticisms depend on conflating old-style extensional interpretations with worlds.
[26] The most elegant and perspicuous account of higher-order intensional type theory is that developed by Pavel Tichý: see his (1981) for a purely formal presentation and his (1978b) for philosophical background.
[27] This paper was written while I was a visitor at the University of Helsinki, and on leave from my position at Otago University. I would like to thank my colleagues at Otago, for taking over my teaching duties; the Finnish Academy, for a research fellowship which made my visit to Helsinki possible; and the philosophers in Helsinki, who provided such a stimulating environment in which to work.

REFERENCES

Armstrong, D. (1978) *A Theory of Universals*, Cambridge University Press.
Armstrong, D. (1983) *What is a Law of Nature?*, Cambridge University Press.
Carnap, R. (1947) *Meaning and Necessity*, University of Chicago Press.
Carnap, R. (1966) *Philosophical Foundations of Physics*, ed. M. Gardner, New York: Basic Books.
Carnap, R. (1971) *Studies in Inductive Logic and Probability*, vol. 1, ed. by Carnap, R. and Jeffrey, R., Berkeley: University of California Press.
Devitt, M. (1979) 'Against Incommensurability', *The Australasian Journal of Philosophy* **57**, 29—50.
English, J. (1978) 'Partial Interpretation and Meaning Change', *The Journal of Philosophy* **75**, 57—76.
Miller, D. (1974) 'Popper's Qualitative Theory of Verisimilitude', *The British Journal for the Philosophy of Science* **27**, 166—77.
Miller, D. (1975) 'The Accuracy of Predictions', *Synthese* **30**, 159—91.
Miller, D. (1978) 'Distance Between Constituents', *Synthese* **38**, 197—212.
Musgrave, A. (1979) 'How to Avoid Incommensurability', *The Logic and Epistemology of Scientific change* (*Acta Philosophica Fennica* **30**, nos. 2—4), 336—46.
Nola, R. (1980) '"Paradigms Lost, or the World Regained" — An Excursion into Realism and Idealism in Science', *Synthese* **45**, 317—350.
Oddie, G. (1981) 'Verisimilitude Reviewed', *The British Journal for the Philosophy of Science* **32**, 237—65.
Oddie, G. (1986) *Likeness to Truth*, Dordrecht: Reidel.
Pearce, D. (1983) 'Truthlikeness and Translation: A Comment on Oddie', *The British Journal for the Philosophy of Science* **34**, 380—5.
Popper, K. (1963) *Conjectures and Refutations*, London: Routledge and Kegan Paul, references to the 1972 edition.
Smart, J. J. C. (1986) 'Realism v. Idealism', *Philosophy* **61**, 295—312.

Tichý, P. (1971) 'Synthetic Components of Infinite Classes of Postulates in *Archiv für Mathematische Logik Und Grundlagenforschung*', **14**, 167—78.
Tichý, P. (1974) 'On Popper's Definitions of Versimilitude', *The British Journal for the Philosophy of Science* **25**, 25—42.
Tichý, P. (1978a) 'Verisimilitude Revisited', *Synthese* **38**, 175—96.
Tichý, P. (1978b) 'Two Kinds of Intensional Logic', *Epistemologia* **1**, 143—64.
Tichý, P. (1981) 'Foundations of Partial Type Theory', *Reports of Mathematical Logic* **14**, 57—72.
Williams, P. (1973) 'On the Logical Relations Between Expressions of Different Theories', *The British Journal for the Philosophy of Science* **24**, 357—408.
Winnie, J. (1975) 'Theoretical Analyticity', *Rudolf Carnap: Logical Empiricist*, ed. J. Hintikka, Dordrecht: Reidel.

GREGORY CURRIE

REALISM IN THE SOCIAL SCIENCES: SOCIAL KINDS AND SOCIAL LAWS

1. NATURAL KINDS AND NATURAL LAWS

A general picture that informs many versions of scientific realism is this. The physical world exists independently of us. In one sense, of course, we physically embodied creatures are part of that world; but we do not create it, our thought and action do not sustain it, and there are structural features of it ('the laws of nature') that we can do nothing to alter. It contains things that we never will observe, that perhaps we could never observe, but about which we may come to have some rational beliefs by way of plausible scientific inference. The aim of science is to construct the theories of ever increasing truth-likeness about what the physical world, in both its observable and unobservable aspects, is like.

This picture contrasts sharply with the various claims of anti-realists: that to accept a scientific theory is always and only to accept that it is empirically correct; that there is no gap between a theory meeting all determinable constraints and its being true; that we can form no coherent conception of the truth conditions for a statement that cannot, even in principle, be verified.[1]

Recent attempts to elaborate the realist picture have emphasized the importance of two connected notions; natural kinds and laws of nature. Natural kinds like water, gold and tiger are said by Hilary Putnam and others to have 'real essences' — inner natures that underly the commonly known traits of things. The source of their identifying characteristics is not our definitional activity but rather features of the world independent of us. These real essences can be investigated, but our evidence about them is indirect and deeply embedded in scientific theorizing. Accepting some particular view about what these essences are involves us in accepting conjectural scientific theories, not merely as empirically adequate, but as true or approximately true accounts of the structure of the physical world.[2]

The concept of a law of nature has recently enjoyed the attention of scientific realists like David Armstrong.[3] Armstrong urges us to

Robert Nola (ed.), Relativism and Realism in Science, 205—227.

reject the traditional empiricist idea of a law as a mere regularity of succession between events; a regularity conventionally but somewhat inadequately expressed in the form '$\forall x(Fx \supset Gx)$'. For empiricists face grave difficulties about how, for example, laws conceived of as mere regularities between kinds *F* and *G* can entail counterfactuals of the form 'If this had been an *F* it would have been a *G*'. It seems that something stronger than regularity is required. Some realists like Jack Smart reject the strong view of laws and prefer an account in terms of regularity. But in doing so I think they compromise their realism. For Smart has to distinguish laws from accidentally true generalizations by saying that geuine laws are regularities derivable within a unified and well corroborated theoretical system; ontologically speaking, laws are nothing more than regularities; what is distinctive about them is how they function for us.[4] Thus the distinction between laws and accidental generalizations becomes a methodological rather than an ontological one. Someone who is a realist about laws ought to say that the distinction is grounded in the nature of things, not merely in our scientific theorizing.

The suggestion is, then, that the full blooded realist may interpret laws as relations of nomological necessitation holding between universals (properties or kinds). Laws, on this conception, entail regularities, but the statement of a law is stronger than the statement of the corresponding regularity; there may be regularity without nomological connection. Such laws sustain counterfactuals, and deliver the intuition that the laws of nature are beyond our powers to affect. Whatever we might have done to create examples of some natural kind, the laws relating kinds would be unaltered. If kinds *F* and *G* stand in the relation of nomological necessitation, then if we had made this thing an *F*, it would also have been a *G*.[5]

Thus on the realist conception of the natural world, two things stand out as distinctively independent of our thought and action; the nature of natural kinds, and the existence of natural laws.

I am interested in the realist's account of both these things, because it seems to me that when we consider what can be said on analogous topics about the social world we come to see the extent to which we are forced to be anti-realists about the social sciences. I do not believe that there can be any 'laws of society' in anything like the sense that there may be laws of nature. And I believe that there is a sense in which social kinds depend both for their natures and their occurrence on our thought and action. Thus the anti-realism I am arguing for does not emerge from a generally anti-realist perspective on knowledge and the world such as is exemplified in the writings of Michael Dummett and Bas van Fraassen.[6] I am arguing that there are reasons peculiar to the social sciences that foreclose on certain realist options. In this sense I am arguing against the 'unity of method' thesis, about which I shall say a little more at the end of this paper.

The results about social kinds and about (putative) social laws that I want to establish here both depend upon a supervenience thesis that I have argued for in an essay called 'Individualism and Global Supervenience'. In the present essay I rely upon an intuitive and briefly argued version of that thesis. For a more formal elaboration, the reader is referred to the earlier essay.[7]

One further preliminary remark. I do not have a precise account of what it takes for something to be an example of a *social* kind. Roughly speaking, I think that a social kind is one the instantiation of which presupposes a system of interlocking beliefs and desires on the part of rational agents. (David-Hillel Ruben gives an account along these lines[8].) However, I do not want to rest what I say in the body of this essay on any particular analysis of social kinds, for I can rely, I think, on a fair degree of consensus as to what particular things are instances of social kinds, even where there is disagreement about how that notion is to be analysed. The things I have in mind are physical objects which have a social function, like items of exchange and the trappings of devotion, social structures such as the family, the state and the working class, institutions such as banks, businesses and the cabinet, 'offices' such as head of state, chairperson of the board, secretary of the club, together with more abstract kinds of things such as languages and other conventional systems like laws and customs. Particular instances of these things are exemplars of social kinds.

2. SUPERVENIENCE

It is evident upon even superficial reflection that the social world is not independent of us in quite the same sense that the physical world is. There could not be a social world without rational agents, and the status of certain things as social objects (e.g. money) depends essentially upon how they are regarded by such agents. But this voluntaristic picture of the social must be tempered by various concessions to the rigidity, resilience, autonomy and independence of social things. What social facts are true is largely independent of what we think and do; and this is true of many social facts about ourselves. I cannot change my social class at will, I cannot change the class I was born into at all. There are social facts about me that I do not know and may never know. There may be social facts about me that nobody knows. I may be exploited in the labour market, whether anybody knows it or not. There are things about our own society that we would very much like to know, and things about it we would very much like to change, but in some cases we may not be able to do either. Vico's model of knowledge in the social sciences — what we create we know with certainty — does not stand up to pressure from these commonsensical examples.[9]

We need a precise characterization of the dependence of the social world on our activities; a characterization that makes room for the limited autonomy to which I have just alluded. The idea of supervenience will help us here. Intuitively, Ψ supervenes on Φ if and only if there can be no variation in Ψ without variation in Φ, where Φ and Ψ are kinds or classes of properties or traits. If Ψ supervenes on Φ then two things similar with respect to the Φ properties will be similar with respect to the Ψ properties. This idea can be made precise in various ways, and different theses of supervenience require different formulations. We face a decision, in particular, about what kinds of things are said to be relevantly similar. Narrowly interpreted, the thesis might focus on particular *individuals* and say that any two individuals (e.g. persons) which are similar with respect to the Φ properties (e.g. physical properties) will be similar with respect to the Ψ properties (e.g. mental properties). Taking a broader view, we might focus on *structures*. Consider a class of individuals c_1 and imagine the Φ properties to be distributed through them; this gives us a structure $\langle c_1, \Phi \rangle$. Suppose it is the case that if $\langle c_1, \Phi \rangle$ and $\langle c_2, \Phi \rangle$ are isomorphic, then $\langle c_1, \Psi \rangle$ and $\langle c_2, \Psi \rangle$ are isomorphic. This is another sense in which Ψ supervenes on Φ. Further broadening is achieved if we turn to *possible worlds*. Consider two worlds ω_1 and ω_2, containing the same individuals. Suppose it is the case that if ω_1 and ω_2 are indistinguishable from one another with respect to the distribution of the Φ properties among individuals, then they will be similarly indistinguishable with respect to the Ψ properties. Here we have a further sense in which Ψ supervenes on Φ. Such a supervenience thesis I shall call a *global* supervenience thesis. I believe that there is a true supervenience thesis of this kind that we can formulate concerning the relation of the social to the individual. (In order to avoid confusion, I shall from now on take 'individuals' to refer to persons.)

There may be other kinds of supervenience theses that relate these two kinds of properties, but a global formulation seems to me attractive for two reasons. These reasons have to do with a certain kind of logical strength and a certain kind of logical weakness that such a formulation has. Such a formulation is *weak* in that it does not say that supervenience can be established as a result of a point by point comparison. It does not say that two individuals who are intrinsically alike with respect to individual properties will be alike with respect to social properties. And that is an advantage because a person's social characteristics depend not only on what he or she does but on what others do as well.(Our tentative definition of a social property in the previous section captures this idea). Such a formulation is *strong* in that it claims that the supervenience relation necessarily holds; for it is the claim that any two worlds that are alike in the distribution of individual traits are alike in the distribution of social traits. And I, at any rate, have the intuition that the dependence of the social on the individual is in some sense a conceptual, necessary truth. It is that sense of necessity that the global formulation is intended to capture. Can we formulate and defend such a supervenience thesis relating social kinds and individuals?

We might try a formulation of this idea amenable to our present purposes by saying that what social kinds are instantiated (money, democracy, capitalism, etc.) is determined by ('is supervenient upon') our beliefs, desires and expectations, in conjunction, perhaps, with our sensory experiences, where these internal states are characterized in a purely qualitative, 'notional', 'narrow' or *de dicto* way, rather than by reference to their objects (*de re*).[10] Imagine a possible world exactly like our own in terms of the internal qualitative states of subjects. Could such a world differ from ours in respect of the social kinds it exemplifies? If it could not, we have a clear disanalogy with the realist's conception of the natural world. For a plausible realist principle says that the qualitative states of all individuals radically underdetermine the nature of the physical world — the kinds and laws that it exemplifies.[11] There are possible worlds like the actual world from the point of view of our internal states, but different in respect of the microstructure of the physical environment.

But this way of locating the difference between the physical and the social environments does not quite work. For the qualitative states of individuals do *not* determine the nature of the social world. In particular, a world exactly like our own from the point of view of the mental may be one in which there is no social environment at all. To see this we have to engage in some rather *outré* thought experimentation, but the thought experiment I have in mind represents a *possible* state of the world, and that is enough.

Imagine that Fred lives in a solipsistic world. He is a pure Cartesian ego, or a brain floating in a vat of nutrient. His internal, purely qualitative mental states are just like mine.[12] The differences between us (hopefully rather radical ones) concern our relation to the outer environment. There are no other people with whom he interacts, since he is the sole inhabitant of his world. I think it will be agreed that Fred, whatever he thinks of the matter, does not live in a social world of any kind. One thing this example shows is how independent the social world is from any particular person's mental life: my own inner mental life does not (logically) guarantee me a social environment at all, so *a fortiori* it does not guarantee that I have a social environment with any particular characteristics. But now suppose that there are two such isolated individuals in our world, Fred and Jim. They each have mental states; and their mental states may even be in harmony with each other to the extent that it seems to them that they inhabit a common world with a degree of regularity and coherence. Their experiences may be just like yours and mine. But there are no substantive connections between Fred and Jim. If Jim were to disappear, Fred's experiences would go on in the same way, and *vice versa*. Now if Fred on his own does not live in a social world, Fred and Jim together do not live in one either. In the first case there is just Fred and the movie he is watching (so to speak). If Jim is watching a qualitatively similar movie, that does not create for them a social environment, where one did not exist before. It doesn't matter how many Cartesian egos or brains in vats you add, you don't get a social world unless you add interaction between the individuals concerned. If I am right about this, there could be a world qualitatively identical to ours in terms of 'narrow' mental states, but totally lacking a social dimension. So narrow psychology, even globally considered, does not determine the social. If realism about the social world consists in saying that what is true about the social world is not a function of our inner mental states, then realism about the social is true.

But this is not the end of the matter, for the anti-realist will point out that we are not comparing like with like. The realist about the physical world thinks that the physical world is not determined by what we think *and* what we do; the physical facts do not supervene on the totality of facts about our *actions* (construing action as having an intentional and a behavioural component). The question to ask about social kinds is whether they, unlike physical kinds, supervene on the totality of actions. I shall come back to social kinds in a moment, after a discussion of natural kinds that will set the scene.

3. SUPERVENIENCE AND NATURAL KINDS

Suppose, with Hilary Putnam, that Earth and Twin Earth are indescernible from the point of view of mental and behavioural states.[13] On Twin Earth there is a community of interacting individuals who think, experience and behave as we do. Each of us on Earth has a counterpart on Twin Earth whose inner mental life and bodily movements are indistinguishable from our own. It is clear that there may still be a great deal of variation between Earth and Twin Earth. It may well be the case that no natural kind instantiated on Earth is instantiated on Twin Earth. That is, while there is a wet, drinkable stuff that fills the lakes and seas of Twin Earth, that stuff is not water, because it is not made of H_2O molecules. It is made out of a quite different stuff, XYZ. This is easy to see if we imagine that Earth and Twin Earth are alike in being pre-scientific societies; societies with no criteria for determining the 'real essences' of the stuff that surround their members, including the drinkable stuff in lakes. If Earth and Twin Earth are scientific societies that have not yet got around to identifying the real essence of that stuff, it may be the case that there is water on Earth and none on Twin Earth, without there being any differences at the purely individual level between them. This may still be true even if both are scientific societies that have identified the inner constitution of that stuff as H_2O; it is just that on Twin Earth they got it wrong; what they think is H_2O is actually XYZ. It is even arguable that in the case imagined there may be no natural kinds instantiated on Twin Earth. Twin Earth is a Berkeleyan world in which there is nothing but appearances. Even if we grant Berkeley's notoriously controversial claim that there really is, in this situation, an external environment for these beings to occupy — grant, in other words, his claim to be affirming common sense — it is not an environment that contains any stuff that has a real essence. For

Berkeley insists, as his position requires of him to do, that there are no hidden essences in a world such as he believed ours to be.

Notice that I am not saying anything here that depends on controversial premises about the meaning of the term 'water'. I am not insisting, with Putnam, that our ordinary use of the term 'water' is responsive to possibly unknown facts about what water is made of, and that, consequently, the meaning of 'water' is not 'in the head'. For we can grant, for the sake of argument, that 'water' as used by a pre-scientific society, and as used by a scientifically ignorant member of a scientific society, is analytically associated with the 'nominal essence' of water, its taste, appearance, typical location and other phenomenal properties. Let us also ignore the fantastical possibility that Twin Earthians are Berkeleyan creatures with no mind-independent physical environment. (Not ignoring it would merely strengthen my case, for if such a thing is a real possibility there would be water on Earth but not on Twin Earth.) Let us grant, in other words, that in the case imagined there is water on both Earth and Twin Earth. But now 'water' has lost its rigidity. It now refers to the stuff, whatever it is, that has those phenomenal properties. On Earth that stuff happens to be H_2O; on Twin Earth it happens to be XYZ. So Earth and Twin Earth differ in that H_2O is exemplified on Earth but not on Twin Earth, while XYZ is exemplified on Twin Earth but not on Earth. On no reasonable assumptions do mental and behavioural states determine what natural kinds there are.

4. COMPARISONS WITHIN A WORLD AND COMPARISONS BETWEEN WORLDS

We must now attend to an issue of modality that I have so far glossed over and which has lent confusion as to the consequences of Twin Earth thought experiments.[14] For the above case to be coherently describable, Earth and Twin Earth must be located in different possible worlds. If Earth and Twin Earth are in the same world, they do not differ as to what natural kinds are instantiated, for both H_2O and XYZ are instantiated in *that* world. They differ only as to what natural kinds are locally available (a point we shall return to). The case I am considering is one in which Twin Earth exists in a possible world distinct from ours, one in which XYZ plays the phenomenal role played in our world by H_2O. (It does not matter, from the point of view of this example, whether we treat Twin Earth as Earth, with us in it, under counterfactual circumstances, or as a counterpart of Earth, with our counterparts in it.[15]).

Cases like this — cases of transworld comparison — are of no interest for Putnam's semantical programme, because he wants to establish that variations in the extension of the term 'water' are consequent upon variation in its intension, where mental states are assumed fixed; and to establish, in consequence, that meaning is not determined by psychological state. For his purposes, Earth and Twin Earth must be located in the same possible world. Comparison across possible worlds does not enable us to infer variation in intension from variation in extension; variation in extension across possible worlds *without* variation in intension is just what we would expect on the conventional theory of meaning. But as I hope the previous section has made plain, I am not arguing for a Putnam style analysis of terms like 'water'. What I am trying to establish is that there is a crucial difference, on any plausible semantical theory, between the ways in which natural kind terms vary in extension across worlds, and the ways in which social kind terms do. For if we now consider how things are on our (counterfactual) Twin Earth when it comes to the instantiation of social kind terms, we see that *things are exactly as they are in the actual world here on Earth*. For it is surely absurd to claim that, while there are no differences between Earth and Twin Earth at the level of individual action, there is yet some difference between them at the level of the social. Suppose, to simplify the issue a little, we assume that Earth is a socially monolithic culture in which capitalism prevails. Could Twin Earth somehow be not a capitalist society? The idea seems incoherent. Without some difference between the two describable in terms of individual thought and behaviour, the attribution of a social difference is surely empty. But, as we have seen, it is uncontentious (for the realist) that counterfactual Twin Earth may differ from Earth as to what natural kinds are instantiated.

Interestingly enough, we get another result about the difference between natural and social kinds if we consider intra- rather than inter-world comparisons. Suppose now that Twin Earth is a distant planet, located somewhere in our space-time continuum, but too far away for us to communicate with. Then we get the result that the social kinds locally available to Twin Earthers are exactly the same as the social kinds locally available to us; no difference in the social kinds a society contains without individual differences. And that, as I have remarked, is not the case with natural kinds. There may be no H_2O that any (actual world) Twin Earther has ever been in contact with, and no XYZ that any of us has ever been in contact with. (If we pass from a consideration of social kinds to a consideration of particular social objects, the same thing cannot be said. On Earth there is France, the Red Cross, the Bank of New Zealand; these are particular social objects. None of these objects exist on Twin Earth, though there they have social objects indistinguishable from them, just as their people are indistinguishable from us. I shall say more about these tokens of social kinds in Section 7.)

5. BETWEEN THE MENTAL AND THE PHYSICAL

Let us now suppose that Putnam's account of natural kinds is roughly correct; that they have real essences not analytically associated with their commonly identifying features, that terms denoting them get their meaning by ostensive confrontation with a sample of the relevant kind, that the use of these terms is subject to division of linguistic labour, etc. In this section I want to examine the question of whether anything similar can be said concerning social kinds, bearing in mind of course the thesis of supervenience that we have enunciated concerning these kinds. Much will depend upon the application and interpretation of Twin Earth cases of the kind already discussed. The result will be to situate social kinds between two extremes; natural and mental kinds.

It is, first of all, no part of my claim that we never make *discoveries* about our social world. Indeed, my thesis is quite consistent with the idea that we are radically mistaken about the nature of our social institutions, that current social theory is radically wrong, and that some other theory will draw the boundaries between social kinds in quite other ways. (Perhaps neoclassical economics and Marxist economics are competitors in this respect.) If this is indeed correct then we may say that social kinds, like natural kinds, have hidden or real essences; that the truely identifying features of social kinds are distinct from those features that we commonly and pre-theoretically employ in order to identify their instances.

We need to notice first of all that Putnam himself thinks that his account of natural kind terms (defined in terms of an unknown hidden essence, subject to the construction of Twin Earth cases, etc.) is applicable to kind terms other than natural kind terms. He does not discuss such terms as 'exchange', 'democracy', 'state', etc. — the sort of terms I am most interested in here — but he does focus on 'artefact' terms like 'pencil' and at least some artefact terms seem to have an essentially social use. (e.g., 'coin', 'telephone', 'radio'). And since Putnam says that his theory applies 'to the great majority of all nouns and to other parts of speech as well'[16], we may assume that what he has to say about artefact terms he intends as applicable to at least some social kind terms. What he says about 'pencil' is actually rather perplexing. He suggests that 'pencil' is just like a natural kind term in that it is not defined in terms of a nominal essence.

When we use the word 'pencil', we intend to refer to whatever has the same *nature* as the normal examples of the local pencils in the actual world.[17]

He then asks us to consider a thought experiment in which pencils (the local ones) are discovered to be organisms.

We cut them open and examine them under the electron microscope, and we see the almost invisible tracery of nerves and other organs. We spy on them, and we see them spawn, and we see the offspring grow into full-grown pencils.[18]

Putnam's point is not that it is possible for pencils to be organisms. It is that there could be things that are the *epistemic counterparts* of pencils; things that seem to us to be pencils but which, because of their organismic natures, are not. If Twin Earth contained such things we would say, according to Putnam, 'the things on Twin Earth that pass for pencils aren't really pencils. They're really a species of organism', just as we would say that what passes on Twin Earth for water is not really *water*, but XYZ.[19] So pencils have hidden essences after all; their natures are not exhausted by the way they appear to be. And that they do so entails that there are Twin Earth cases for pencils; things are the same on Twin Earth as they are on Earth when it comes to how people think and behave, but there might be no pencils there. Obviously, Putnam's argument has as much, or as little, plausibility if we substitute social kinds like *coin* or *telephone* for *pencil.*

In fact, Putnam's conclusion seems unbelievable. We would *not* deny that these otherwordly things were pencils simply because they were organisms; if they were used just as we use pencils they would *be* pencils. Putnam seems to have forgotten at this point an important ingredient in his own theory; that what counts as the real essence of a kind differs from context to context. For a species it is genetic structure that counts; for a substance like water it is molecular structure; for acids it is being a proton donor. Confronted with a sample of *aqua fortis*, the question, 'to what kind does that belong?' is ambiguous. If we are concerned with molecular structure, the answer is 'HNO_3'; if we are concerned with certain sub-atomic characteristics, the answer may be 'acid'. Even if 'pencil' functions in a way that fits Putnam's account of natural kind terms, the mere fact that all local pencils are organisms does not establish that pencils are necessarily organisms, anymore than all local acids being HNO_3 would establish that all acids are necessarily HNO_3.

Perhaps Putnam's example establishes one thing (and I think it is intended to establish at least this); that so called 'artefact' terms do not necessarily refer to artefacts – i.e. things made by us. But I don't want to worry about this. Let us grant that it is not analytic of 'pencil' that pencils are artefacts. What I do insist upon is that pencils (or coins, or telephones) are necessarily *functional* objects of some kind; their natures are specified in terms of the purposes to which we put them. In saying this I am still saying something that Putnam must deny. (This is important because many social kinds are such that we identify them, or their instances, in terms of their functions. Obvious examples are units of economic analysis like firms, households, banks and discount houses.)

To say this is, of course, perfectly consistent with Putnam's general views about kind terms, as Hilary Kornblith points out.[20] The story we would then have to tell about terms like 'pencil' is this. The term 'pencil' is introduced in the following way. Someone says 'this is a pencil, and anything with the same function as this is a pencil too'. Now with pencils this is not a very plausible story, since it is hard to imagine a state of wide-spread or universal ignorance about what pencils are for. And if everybody knows what a pencil is for, then the term 'pencil' is likely to be introduced as a shorthand for some description of that function. But cases that fit Putnam's model are at least imaginable. Some artefact terms do seem to be associated with a division of linguistic labour. My knowledge of the identifying conditions for a transformer is as abysmal as my knowledge of the identifying conditions for elm trees. This account is also plausible for some of the terms I am particularly interested in here. Our use of the term 'democracy' may well be consistent with a story to the effect that a democracy is a society that functions in the same way that Fifth Century Athenian society did, whatever that is. And I certainly defer to experts when it comes to applying concepts like 'profit maximization', 'alienation' and 'exploitation'.

So let us grant then that terms denoting social kinds generally function as if they had been introduced by reference to a paradigm example, together with the stipulation that anything else is a social entity of that kind if it has the same function. Even if we grant this, the following result holds: if counterfactual Twin Earth is like Earth in respect of individual thought and action, the same social kinds will be found there as here. There may be ignorance or error, here as on Twin Earth, as to the function that the members of some social kind have. But whatever the function of a social institution or other kind of social entity is here on Earth, the corresponding entity on Twin Earth must surely have the same function. If it turned out to be the case that the things they use on Twin Earth as telephones were organisms, even though nobody on Twin Earth ever came to know this, those things would still be telephones. The only reasonable grounds for saying that they are not telephones would be that they are used in some way that differs from the way they are used here on Earth; and that is ruled out by the stipulation that Earth and Twin Earth do not differ from the point of view of the actions of individuals. And if there is profit maximization (whatever that is) here on Earth, there is profit maximization on Twin Earth too. So even if social kind terms were like natural kind terms in admitting division of linguistic labour and definition in terms of a paradigm and an equivalence relation, they still would differ from natural kinds in that their distribution could not vary independently of variations in human activity. To that extent they fail to be independent of us; to that extent we cannot be realists about them.

I have said that social kinds may have real essences, and that we cannot construct Twin Earth cases for them. This may sound like a paradoxical result, for the possibility of Twin Earth cases is sometimes taken to be a test for whether a kind has a real or a nominal essence. But the air of paradox is dispelled when we notice the very strong demands that I place upon the construction of Twin Earth cases. I require that Twin Earth be a place that is, from an individual point of view, qualitatively indistinguishable from Earth. But sometimes Twin Earth is not thought of in this way. Colin McGinn, for instance, in discussing the difference between physical and mental kinds, supposes

Twin Earth to be a place which differs considerably from Earth in individualistic respects, but which is similar to Earth in this: that both communities employ the same commonly acknowledged criteria for determining the application of mental terms.[21] McGinn asks: could what Twin Earthers call 'pain' be a different kind of thing from what we call 'pain'? And his answer is 'No': terms denoting mental kinds acquire their meanings within the 'folk psychology' that sums up the criteria for determining the application of these terms. Same criteria, same folk psychology, same meanings for homophonic mental kind terms. If there is pain on Earth, there is pain also on Twin Earth, and similarly for all other mental kinds. It is not possible to construct Twin Earth cases (in this sense) for mental kinds. And mental kinds have no real essences, nor are they subject to division of linguistic labour, nor is their application subject to revision or overthrow through scientific investigation.

Given what I have said already, it is plain that McGinn-type Twin Earth cases are constructible for at least some social kind terms. What social kinds there are depends upon the totality of what people do and think, but not on that more restricted part of thought and behaviour that constitutes the ways we commonly acknowledge for identifying social kinds. We and the Twin Earthers may be similar in our choice of criteria, but radically different in other ways. And these differences may be such that, though we agree about how to identify exploitation, that social kind is instantiated here but not there.

McGinn's view of mental kind terms is controversial (it is controverted by writers like Churchland and Stitch who believe that folk psychology is false[22]). Notice, however, that it would not be in the least controversial to claim that mental kind terms are immune to the construction of Twin Earth cases in my sense. For it is trivial that if two worlds do not differ in any individual respect (behavioural or mental) they do not differ in respect of what mental kinds are instantiated within them. Suppose that McGinn (and others, like David Lewis[23]) are right in claiming that folk psychology is immune to substantial revision. Then there is an interesting sense in which social kinds occupy a place intermediate between physical and mental kinds. Call the kind of Twin Earth case that I have exploited in this paper *strong*. Call the kind of Twin Earth case that McGinn employs *weak*. Then we may say: physical kinds are subject to both strong and weak Twin Earth cases; mental kinds to neither. Social kinds are subject to weak Twin Earth cases but not to strong ones. If Twin Earth cases are a test of whether a realist view of a certain class of kinds is possible, realism about kinds admits of degree.

6. LAWS OF NATURE AND LAWS OF SOCIETY

I turn now to a consideration of laws in the social sphere. Given our realistic conception of law, the problem is to decide whether the existence of an observed regularity of connection between social kinds or predicates could ever provide the same sort of evidence for the existence of a social law that similar observed regularities in the natural world are taken to provide. For it is clear from what I said at the beginning of this essay that a mere regularity (however widespread) at the social level will not constitute a law, though of course the presence of such a regularity may suggest that such a law exists. We do not want to know whether universal regularities may exist in social life (clearly they may), but rather whether the *best explanation* of such regularities can ever be that the appropriate social kinds or properties stand in some nomological relation (that may be probabilistic rather than strictly determining).

Notice that in other cases of supervenient phenomena — ethical, aesthetic, semantical and, more contentiously, mental phenomena — we do not expect there to be lawlike connections between properties characteristic of those phenomena. We have, of course, chemical laws, and chemistry supervenes on physics; no chemical difference without a physical difference. But here we have reducibility as well as supervenience. Chemical laws are, in principle, restatable as laws of physics. But the relation between social and individual kinds (mental and action kinds) precludes, I think, reducibility. We cannot give necessary and sufficient conditions for the application of a social predicate in terms of the applicability of individualistic predicates. Social concepts display individual plasticity in the way that mental ones display neural plasticity (more on this in Section 7). In general, where we have supervenience without reducibility, we do not expect to find causal laws of the supervenient phenomena.

Could the social world be some exception to this? I believe not. On the view of laws that we have been developing here, a law is, in Hume's terms, a 'distinct existent'; we cannot deductively infer the existence of a law from other phenomena. (Of course if we believe that there are derived laws — laws that hold by virtue of other laws that hold — these laws will not be distinct existences in our favoured sense. As it will become clear, I shall consider only basic, non-derived laws here.) In particular we cannot infer the existence of a law from a regularity — an observed one or a conjectured universal one. Let us abbreviate the proposition that F and G are regularly conjoined as 'Reg(F, G)', and the (stronger) proposition that F and G stand in the relation of nomological necessitation as 'Nom(F, G)'. Now Nom$(F, G) \models$ Reg(F, G), but not vice versa.[24] Laws of nature, on the realist conception, are 'logically subtractable' elements in the world. Suppose our world is one in which there are natural laws, including one to the effect that Nom(F, G). Then there is another world like our own except in that there are no laws in that world but only regularities, including Reg(F, G). Such a 'Humean counterpart' world does not differ from the actual world with respect to any particular matter of fact other than Reg(F, G), together with whatever other laws prevail in the actual world.[25]

At this point I want to make an undefended assumption, though the assumption is both reasonable and widely held: there are no laws governing human action. No laws, for example, to the effect that someone in mental state M will perform bodily movement N. In general there are no laws that relate intentional mental predicates to one another or to predicates descriptive of bodily movement.[26] At the level of individual description, there are only particular matters of fact (and their quantificational generalizations) concerning the thoughts, sensations and behaviour of agents. If this is correct, any social laws there are must be *sui generis*; they cannot be consequent upon laws about individuals.[27]

Now assume that there is such a law that holds in a world ω_1 — a law relating social kinds. Such a basic law must be a distinct existence. Hence there must be a world ω_2 which is a Humean counterpart to ω_1 in that it differs from ω_1 only in that our law does not hold in ω_2. ω_1 and ω_2 are the same, therefore, from the point of view of the mental and behavioural states of individuals. But this cannot be. For the social supervenes on the individual; no difference at the level of the social without a difference at the level of the individual. Social laws, construed realistically, cannot meet the requirement that they be distinct existences. But that is a defining condition on a law realistically construed. Hence there can be no social laws.

More generally, we have the following result. If B supervenes on A and there are no laws of A phenomena, then there are no laws of B

phenomena either. We also have confirmation for our intuition that there cannot be ethical, aesthetic and semantic laws. For it is natural to suppose the ethical, aesthetic and semantic facts supervene on facts about individual action in the same way that social facts do. In that case the same form of reasoning that I have just given can be applied to show that there are no such ethical, aesthetic or semantical laws.

Accepting this view about the impossibility of social laws does not preclude our ever having a rationally based expectation that a social regularity will continue. We may have such a reason if we think that the social regularity is the product of some regularity in the behaviour of individuals, and for many such regularities we have every reason to expect that they will continue; because, for instance, the regularity is part of a deeply entrenched and/or useful convention, or merely because it would be, for whatever reason, difficult for people to alter their behaviour in the relevant respects. We may even be able to formulate intuitively correct counterfactual claims of the form 'If this had been an *F* it would have been a *G*', where '*F*' and '*G*' are social kind terms. But our reasoning about such counterfactuals crucially involves the perhaps tacit assumption that, in the imagined counterfactual situation, individual behaviour will be sufficiently close to actual behaviour to sustain, by way of the supervenience relation, the regularity. This kind of reasoning is vague, and perhaps irremediably so, but so is a good deal of our counterfactual reasoning about other matters.

7. REDUCTIONISM

In asserting the supervenience of social kinds on thought and action I have not, of course, been arguing for the reducibility of social kinds. It is very implausible to suppose that they are reducible in any methodologically relevant sense.[28] First of all, certain kinds of social institutions must be explained in terms of their relations to other such institutions (banks lend money). There is no reason to think that there is a convenient hierarchy of such kinds, with those at base level explainable in terms of their relations to individuals. Secondly, an explanation of social institutions in terms of, say, attitudes, would require recourse to unreduced social concepts in specifying the content clauses of those very attitudes.[29] In that case we have, as I remarked in Section 6, the kind of supervenience without reducibility that leads us to affirm that ethical, aesthetic and semantical facts are *sui generis*. We can no more eliminate talk about social kinds than we can eliminate talk of the right, the beautiful and the true.

The possibility remains, however, that some kind of token-token reduction of the social to the individual may be possible, perhaps along the lines suggested by those, like Davidson and Fodor, who think that while mental kinds are not identical to physical kinds, mental particulars (e.g. event tokens) are identical with physical particulars.[30] For tokens of some social kinds reductive identification seems plausible and even necessary. Every token of the kind *money* is a physical object; every token of the kind *head of state* is an individual person. If Lewis is anything like correct in his analysis of conventions, these are regularities of behaviour governed by mutual beliefs.[31] The more controversial cases for reductionism are structures like banks, governments and states. It seems to me that the most plausible reductive strategy for these cases would be to follow Dennett's suggestion for a 'homuncular functionalist' theory of mind. Dennett's idea is that we analyse the mind as a hierarchy of sub-functional homunculi, carrying out tasks of varying degrees of complexity, requiring varying degrees of intelligence.[32] These we analyse in turn into sub-sub-functional components. If our analysis peters out with the postulation of 'stupid' homunculi that carry out purely mechanical tasks, we have a viable materialistic theory of the mind, for the lowest level homunculi are realizable by the hardware of the brain. Similarly if we can explain the functioning of social institutions in terms of hierarchies with individuals at the lowest levels, the prospect opens up for an individualistic theory of the tokens of social kinds. If this is the right approach — and I think something like it must be right — then a strongly realistic theory of such tokens is precluded. By a strongly realistic theory I mean one which says that such tokens exist and are not reducible to anything more fundamental. Such a strongly realistic theory is part of the scientific realist's picture of natural kinds. Tokens of natural kinds are not reducible to things that are not themselves tokens of natural kinds (though many will be reducible to tokens of more basic natural kinds.) This is another respect in which we cannot be realists about the social, in the way that we may be realists about the physical.

8. METHODOLOGY

I said at the beginning of this essay that I was arguing against the unity of method thesis. But I have avoided giving a directly methodological formulation to any of the problems I have dealt with here. I do not doubt that the differences I have argued for between the social world and the physical world have implications for our methodology. In particular, if there cannot be any social laws in the same sense that there may be physical laws, it cannot be the aim of social inquiry to search for them. Of course we do not know that there are laws of nature. We may live in a world of Humean regularities, or in a world with no universal regularities at all. But surely there is nothing conceptually wrong with the idea of a law of nature, and the inference to laws does seem to be inference to the best explanation. But if there are reasons for thinking that there *cannot* be laws of society, we have *a priori* grounds for thinking that social science will go wrong if it devotes itself to the search for such laws. Nor, in that case, can we expect the social sciences to conform to a model of testing and/or justification designed for disciplines where the search for laws is a primary goal. We cannot, for instance, demand that the social sciences formulate falsifiable theories about what the laws of society are if we do not believe that there are such laws. We may want to explain such social regularities as we find, but if we do not believe they are the product of underlying nomological connections, we can have no grounds for regarding them as exceptionless, or even as governed by precisely statable relations of probability.

I defer an examination of how these methodological differences will work out in detail. But one piece of advice naturally suggests itself on the basis of what has been said in this paper. And the advice is that, in searching for a correct methodology of the social sciences, we should turn for our explanatory models to those other areas of inquiry where the phenomena under investigation are recognized to be supervenient in something like the sense of this paper. We may look, as I have done in the previous section, to models of the relation between mental and physical states to guide us in the construction of theories about the relation between social and individual facts. Ethical, aesthetic and semantical analogies may also prove fruitful. Social theorists have tried hard to conform to the positivist's insistence that the theory of society should be modeled on physical theory. This strategy has not, I believe, proved a very rewarding one. Perhaps it is time to look in another direction.

One thing at least is clear: before we can decide on the appropriate methodology for the social sciences we must do a good deal of *a priori* reflection on the nature of social concepts. It is this project that the present paper is intended to advance.

ACKNOWLEDGEMENT

In thinking through the argument of this paper I have benefited from discussions with Graham Oddie and David-Hillel Ruben.[33]

University of Otago

NOTES

[1] See van Fraassen (1980), Putnam (1980) and Dummett (1982) respectively for these positions.
[2] See, e.g., Chapters 8, 11 and 12 of Putnam (1975).
[3] See Armstrong (1978) and (1983), Dretske (1977) and Tooley (1977).
[4] See, e.g., Smart (1985).
[5] As Graham Oddie pointed out to me, Armstrong's version of the theory makes an unwonted concession to anti-realism. For Armstrong insists that there are no uninstantiated laws; laws that relate uninstantiated universals [see, e.g., Armstrong (1983), Chapter 8]. But suppose that it is in our power, as it might be, to produce examples of the uninstantiated *F*. Then it might be in our power to create a law, the law relating *F* and some other universal *G*. But for the realist, what laws there are should be a matter independent of our capacities. I shall ignore the requirement of instantiation in my subsequent discussion of laws.
[6] See, e.g., Dummett (1982) and van Fraassen (1980).
[7] See Currie (1984).
[8] See Ruben (1985), and the discussion of this idea in my review, Currie (1986).
[9] For Vico's idea of 'maker's knowledge' see Berlin (1976).
[10] There is a vast and growing literature on how this is to be done and on whether, indeed, it can be done at all. The *locus classicus* of the debate is Fodor (1980). On *de re* belief see Burge (1977). On wide *vs.* narrow psychology and its relation to realism see Devitt (1984). On the subject's notional world see Dennett (1982). Burge's recently expressed doubts about the possibility of narrow psychology seem to me really to be doubts to the effect that propositions can be the contents of such narrow states, rather than to the effect that the concept of a narrow state is incoherent [see Burge (1982), pp. 114–6].
[11] Geoffrey Hellman gives this as a realist principle in his (1983), p. 236.
[12] Hilary Putnam has recently argued that such thought experiments are incoherent [see Putnam (1981), Chapter 1]. But for the reasons described in Tichý (1986) I reject Putnam's argument.
[13] See Putnam (1975), p. 223.
[14] A point made by Harold Noonan in his (1984).
[15] For counterpart theory see Lewis (1973).

[16] Putnam (1975), p. 242.
[17] *Ibid.*, p. 243. Italics in the original.
[18] Putnam *ibid.*, p. 242.
[19] See *ibid.*, p. 243.
[20] See Kornblith (1980). Kim Sterelny drew my attention to this paper.
[21] See McGinn (1978) and (1980).
[22] See Churchland (1981) and Stitch (1983).
[23] See Lewis (1972).
[24] For the sake of simplicity I assume that the relation Nom(F, G) is determinate in this case rather than probabilistic. Only then do we have Nom$(F, G) \models$ Reg(F, G). On this conception of law, the inductive gap between data and the hypothesis that Nom(F, G) is always wider than the gap between data and Reg(F, G). Why prefer the more risky strategy? Perhaps the answer is that the potential explanatory gains are greater. The claim that F and G are causally related seems to be a better explanation of why *Fs* and *Gs* have been observed to be correlated, than does the claim that F and G are, as a matter of fact, always correlated. [See Foster (1982–3).]
[25] I say 'particular matters of fact' to exclude counterfactuals. For if the realist is correct in thinking that regularities do not entail counterfactuals, subtracting the law Nom(F, G) means also subtracting its counterfactual consequences. But counterfactuals are not particular matters of fact. John Earman denies the strongly realist view of laws when he asserts that 'For any W_1 and W_2, if W_1 and W_2 agree on all occurent facts [i.e. particular matters of fact], then W_1 and W_2 agree on all laws' [Earman (1984), p. 195].
[26] See, e.g., Davidson (1970) and McGinn (1980).
[27] Popper is one philosopher who believes in the existence of *sui generis* sociological laws. He describes the view that social laws are derived from the laws of individual psychology as 'psychologism'. See Popper (1965), p. 88 and p. 332, n. 13.
[28] The exact relation of supervenience to reduction has yet to be explored. My own feeling — corroborated by a reading of John Bacon's [1986] — is that if one is prepared to be very liberal in formulating closure conditions for kinds of properties even global supervenience theses will entail necessary coextensiveness (identity?) between between Φ and Ψ properties. One might allow, for instance, that possibly infinite Boolean combinations of Φ properties result in Φ properties. I do not now want to consider the question of whether possibly infinite disjunctions of possibly infinite conjunctions of individual properties will be individual properties. When I deny that supervenience entails reducibility I mean to deny that it entails reducibility in any form that we could capture in a manageable set of definitions and bridging laws.
[29] See Ruben (1985), chapter 3.
[30] See Davidson (1970) and Fodor (1976).
[31] See Lewis (1969).
[32] See Dennett (1978).
[33] I am also grateful to an anonymous referee for a number of suggestions.

REFERENCES

Armstrong, D. M. (1978) *Universals and Scientific Realism*, Cambridge University Press.

Armstrong, D. M. (1983) *What is a Law of Nature?*, Cambridge University Press.
Bacon, J. (1986) 'Supervenience, Necessary Coextension, and Reducibility', *Philosophical Studies* **49**, 163–76.
Berlin, I. (1976) *Vico and Herder*, London: The Hogarth Press.
Burge, T. (1977) 'Belief *De Re*', *Journal of Philosophy* **74**, 338–62.
Burge, T. (1982) 'Other Bodies', in Woodfield, A. (ed.), *Thought and Object*, Oxford: Clarendon Press.
Currie, G. (1984) 'Individualism and Global Supervenience', *British Journal for the Philosophy of Science* **35**, 345–58.
Currie, G. (1986) Review of Ruben (1985), *British Journal for the Philosophy of Science* **38**, 127–32.
Churchland, P. (1981) 'Eliminative Materialism and the Propositional Attitudes', *Journal of Philosophy* **78**, 67–90.
Davidson, D. (1970) 'Mental Events', in *Essays on Actions and Events*, Oxford: Basil Blackwell, 1980.
Dennett, D. (1978) 'Towards a Cognitive Theory of Consciousness', in *Brainstorms*, Cambridge, Mass.: Bradford Books.
Dennett, D. (1982) 'Beyond Belief, In Woodfield (ed.): *Thought and Object*, Oxford: Clarendon Press.
Devitt, M. (1984) *Realism and Truth*, Oxford: Basil Blackwell.
Dretske, F. (1977) 'Laws of Nature', *Philosophy of Science* **44**, 248–68.
Dummett, M. (1982) 'Realism', *Synthese* **52**, 55–112.
Earman, J. (1984) 'Laws of Nature: Empiricist Challenge', in Bogdan, R. (ed.), *D. M. Armstrong*, Dordrecht: D. Reidel.
Fodor, J. (1976) *The Language of Thought*, Brighton: Harvester Press.
Fodor, J. (1980) 'Methodological Solipsism Considered as a Research Strategy in Cognitive Psychology', in *Representations*. Brighton: Harvester Press, 1981.
Foster, J. (1982–3) 'Induction, Explanation and Natural Necessity', *Proceedings of the Aristotelian Society* **83**, 87–102.
Hellman, G. (1983) 'Realist Principles', *Philosophy of Science* **50**, 227–49.
Kornblith, H. (1980) 'Referring to Artifacts', *Philosophical Review* **89**, 109–114.
Lewis, D. K. (1969) *Convention*, Oxford: Basil Blackwell.
Lewis, D. K. (1972) 'Psychophysical and Theoretical Identifications', *Australasian Journal of Philosophy* **50**, 2249–58.
Lewis, D. K. (1973) *Counterfactuals*, Oxford: Basil Blackwell.
McGinn, C. (1978) 'Mental States, Natural Kinds and Psychophysical Laws', *Aristotelian Society Proceedings, Supplementary Volume* **52**, 195–220.
McGinn C. (1980) 'Philosophical Materialism', *Synthese* **44**, 173–206.
Noonan, H. (1984) 'Fregean Thoughts', *Philosophical Quarterly* **34**, 205–24.
Popper, K. R. (1965) *The Open Society and its Enemies*, 5th Edition, London: Routledge and Kegan Paul.
Putnam, H. (1975) *Mind, Language, and Reality: Philosophical Papers, Volume 2*, Cambridge University Press.
Putnam, H. (1980) 'Models and Reality', *Journal of Symbolic Logic* **45**, 464–82.
Putnam, H. (1981) *Reason, Truth and History*, Cambridge University Press.
Ruben, D-H. (1985) *The Metaphysics of the Social World*, London: Routledge and Kegan Paul.

Smart, J. J. C. (1985) 'Laws of Nature and Cosmic Coincidences', *Philosophical Quarterly* **35**, 272–80.

Stitch, S. (1983) *From Folk Psychology to Cognitive Science: the Case Against Belief*, Cambridge, Mass.: Bradford Books, MIT Press.

Tichý, P. (1986) 'Putnam on Brains in a Vat', *Philosophia* **16**, 137–46.

Tooley, M. (1977) 'The Nature of Laws', *Canadian Journal of Philosophy* **7**, 667–98.

Van Fraassen, B. (1980) *The Scientific Image*, Oxford: Clarendon Press.

ALAN MUSGRAVE

THE ULTIMATE ARGUMENT FOR SCIENTIFIC REALISM

Realism and relativism stand opposed. This much is apparent if we consider no more than the realist *aim* for science. The aim of science, realists tell us, is to have true theories about the world, where 'true' is understood in the classical correspondence sense. And this seems immediately to presuppose that at least some forms of relativism are mistaken. The truth which realists aim for is absolute or objective, rather than relative to 'conceptual scheme' or 'paradigm' or 'world-view' or anything else. And the world which realists seek the truth about is similarly independent of 'conceptual scheme' or 'paradigm' or 'world-view' or anything else. If realism is correct, then relativism (or some versions of it) is incorrect.

But is realism correct? As it stands, this question is ill-defined because realism itself is ill-defined. Obviously, there is more to scientific realism than a statement about the aim of science. Yet what more there is to it is a matter of some dispute among the realists themselves. Whether or not realism is correct depends crucially upon what we take realism to assert, over and above the minimal claim about the aim of science.

My way into these issues is through what has come to be called the 'Ultimate Argument for Scientific Realism'.[1] The slogan is Hilary Putnam's: "Realism is the only philosophy that does not make the success of science a miracle".[2] Slogans are not arguments, and the first task is to find out exactly what this Ultimate Argument is. Surprisingly, this is not an easy task. Clarifying the argument will simultaneously clarify what the realism is for which it is an argument. And then we must of course ask whether the argument is a good argument.

As Putnam's slogan already makes clear, the argument appeals to the (alleged) *success* of science. Such appeals are nothing new: Clavius, Kepler and Whewell made them long before Popper, Smart, Putnam or Boyd. The early appeals were meant to show that the realist aim for science had been *achieved*. Thus Clavius argued that the predictive success of Ptolemaic astronomy showed that the theory was true and that its 'theoretical entities' (eccentrics and epicycles) really existed:

Robert Nola (ed.), Relativism and Realism in Science, 229—252.

> But by the assumption of Eccentric and Epicyclic spheres not only are all the appearances already known accounted for, but also future phenomena are predicted, the time of which is altogether unknown. . . . it is incredible that we force the heavens to obey the figments of our own minds, and to move as we will, or in accordance with our principles (but we seem to force them, if the Eccentrics and Epicycles are figments, as our adversaries will have it).[3]

Clavius was wrong. Eccentrics and epicycles were figments of the Ptolemaic astronomer's imagination. The predictive success of a theory does not entail that it is true or that its theoretical entities really exist. Clavius simply committed the fallacy of affirming the consequent.

Now Clavius was actually aware of this ancient sceptical objection — but he had nothing but hot air to offer against it:

> Next if it is not right to conclude from the appearances that eccentrics and epicycles exist in the heavens, because a true conclusion can be drawn from false premises, then the whole of natural philosophy is doomed . . . all the natural principles discovered by philosophers will be destroyed. Since this is absurd, it is wrong to suppose that the force and weight of our argument is weakened by our opponents. It can also be said that the rule that truth follows from falsehood is irrelevant.[4]

Obviously, Clavius tried to prove too much. So did Kepler when he said that a habitual liar will always be found out, and that a lot of predictive success must establish truth. A string of fallacies does not add up to a valid argument.[5] So did Galileo when he said that the earth must move because postulating that it does explains the tides. So, finally, did Whewell when he said (if he did say it) that predictive success in the form of a 'consilience of inductions' proves truth.

The most that a realist can say is that predictive success yields (inconclusive) evidence for the truth of theory and that such evidence might sometimes make it reasonable to presume that a theory is true and that its theoretical entities really exist. The realist can add, in support of the last point, that it may be reasonable to presume true what subsequently turns out to be false — so that it might have been reasonable for Clavius tentatively to presume that Ptolemaic astronomy was true and the eccentrics and epicycles real.

But realists are not the only philosophers who value predictive success — nor are they the philosophers who value it most. Instrumentalists will say that predictive success gives us (inconclusive) reason to think we have an efficient theoretical instrument of prediction. Van Fraassen's constructive empiricist will say that predictive success gives us (inconclusive) reason to think we have an empirically adequate theory. Laudan's problem-solver will say that predictive success gives us (inconclusive) reason to think that we have a theory which is a good empirical problem-solver. Anti-realists value predictive success as much, if not more, than realists, and can make similar epistemological use of it. So far, then, we have no argument for scientific realism.

Laudan thinks that this is the end of the matter: modern realists simply commit the fallacy of affirming the consequent all over again. He talks of '*The Realist's ultimate Petitio Principii*' as follows:

> It is time to step back a moment from the details of the realists' argument to look at its general strategy. Fundamentally, the realist is utilizing . . . an abductive inference which proceeds from the success of science to the conclusion that science is approximately true, verisimilar, or referential (or any combination of these). . . .
>
> It is little short of remarkable that realists would imagine that their critics would find the argument compelling. . . . ever since antiquity critics of epistemic realism have based their scepticism upon a deep-rooted conviction that the fallacy of affirming the consequent is indeed fallacious.[6]

Quite so. But has Laudan correctly construed the Ultimate Argument as the 'ultimate *Petitio Principii*' of affirming the consequent? I think not.[7]

Before I say how the Ultimate Argument is to be construed, there is another point to be made about Clavius's argument. It concerns his view that the predictive success of Ptolemaic astronomy would be 'incredible' if that theory were not true. This can simply be denied. After all, Babylonian astronomers detected periodicities in astronomical phenomena and devised algebraic rules for predicting them. It is hardly incredible or miraculous that a rule expressly devised to capture some periodic phenomenon should successfully predict future instances of that periodic phenomenon. (What might be said to be incredible or miraculous is that eclipses are periodic phenomena, not that we can devise a rule to capture this. Except that miracles are commonly thought of as *violations* of general laws of nature, rather than as the obtaining of those laws!) Nobody thinks that the Babylonian algebraic rules truly describe some hidden reality. Now if Hellenic astronomers (including Ptolemy) devised geometrical models rather than algebraic rules to accomplish the same predictive tasks, it would hardly be incredible that those models successfully predicted future instances of periodic phenomena such as eclipses.

But what if a theory designed to accommodate one phenomenal

regularity (or set of them) should successfully predict a quite different regularity (or set of them)? That *would* be surprising. A conceptual tool designed to do one job turns out to do a quite different job equally well, a 'figment' dreamt up for one purpose turns out to be well-adapted to a different purpose. It would be as if a plane designed for smoothing wood proved capable of remote tuning a TV set! (I owe the last sentence to Homer Le Grand.)

Hence careful realists, beginning with William Whewell, distinguished two kinds of predictive success, predicting known effects and predicting novel effects. Whewell claimed that no theory which had enjoyed novel predictive success had ever subsequently been abandoned. He seems to have thought that novel predictive success provides *conclusive* evidence for the truth of the theory:

> No accident could have given rise to such an extraordinary coincidence. No false supposition could, after being adjusted to one class of phenomena, exactly represent a different class, where the agreement was unforeseen and uncontemplated.[8]

Again, Whewell's view is too strong. The principle "If a theory has novel predictive success, then it is true" still falls foul of the fallacy of affirming the consequent. A weaker view than Whewell's would be that novel predictive success gives us the best kind of evidence for truth. And a better principle than Whewell's would be: "If a theory has novel predictive success, then it is reasonable to presume (tentatively) that it is true".

All of this depends, of course, on our being able to make good the intuitive distinction between prediction and novel prediction. Several competing accounts of when a prediction is a novel prediction for a theory have been produced. The one I favour, due to Elie Zahar and John Worrall, says that a predicted fact is a novel fact for a theory if it was not used to construct that theory — where a fact is used to construct a theory if it figures in the premises from which that theory was deduced. But this is not the place to elaborate or defend that view.[9]

Popper also draws attention to Whewell's distinction, but makes a quite different point with it:

> There is an important distinction . . . between two kinds of scientific prediction, . . . the prediction of *events of a kind which is known* . . . and . . . the prediction of *new kinds of events* . . . It seems to me clear that instrumentalism can account only for the first kind of prediction: if theories are instruments for prediction, then we must assume that their purpose must be determined in advance, as with other instruments. Predictions of the second kind can be fully understood only as discoveries.[10]

Here the argument seems to be that 'instrumentalism' cannot *account for* or *explain* novel predictive success, whereas scientific realism can account for or explain this. Novel predictive success is not a premise from which we argue *to* something (truth, presumed truth, or whatever). Rather it is a conclusion, an *explanandum*, of which scientific realism is to be part of the *explanans*.

Before analysing this argument any further, it is worth noting that Duhem, Popper's arch-instrumentalist, had already acknowledged that arguments like Popper's *and* Whewell's had some force. They actually led him to spice his (alleged) instrumentalism with a whiff of realism. Duhem writes:

> . . . the consequences that can be drawn from [a theory] are unlimited in number; we can, then, draw some consequences which do not correspond to any of the experimental laws previously known, and which simply represent possible experimental laws . . .
>
> Now, on the occasion when we confront the [novel] predictions of the theory with reality, suppose we have to bet for or against the theory; on which side shall we lay our wager? If the theory is a purely artificial system, . . . if the theory fails to hint at any reflection of the real relations among the invisible realities, we shall think that . . . [we] will fail to confirm a new law. [That we should] would be a marvelous feat of chance. It would be folly for us to risk a bet on this sort of expectation.
>
> If, on the contrary, we recognise in the theory a natural classification, if we feel that its principles express profound and real relations among things, we shall not be surprised to see its consequences anticipating experience and stimulating the discovery of new laws; we shall bet fearlessly in its favour.
>
> The highest test, therefore, of our holding a classification a natural one is to ask it to indicate in advance things which the future alone will reveal. And when the experiment is made and confirms the predictions obtained from the theory, we feel strengthened in our conviction that the relations established by our reason among abstract notions truly correspond to relations among things.[11]

Here Duhem operates, not with realist notions of truth and falsity, but with the notion that some theories are 'purely artificial systems' and others 'natural classifications'. It is not easy to explain how a 'natural classification' differs from a true theory, especially when we are told that in a natural classification "the relations . . . among abstract notions truly correspond to relations among things". No matter. Let us grant that a theory can be a natural classification without being true, and that Duhem gives us only a whiff of realism rather than realism proper. Still, he seems to be saying two things. First, that the highest test which yields the best evidence that we have a 'natural classification' is a successful test of a novel prediction. Second, that only if we think we have a 'natural classification' will we regard successful novel prediction as anything more than 'a marvelous feat of chance'. It is the second point which bears upon the Ultimate Argument.

According to Whewell, Duhem and Popper, then, what is really surprising or miraculous about science, what really needs explaining, is *novel* predictive success rather than predictive success *simpliciter.* I dwell on the point because it is notable by its absence from recent discussions of the Ultimate Argument, by both defenders of the argument (such as Putnam and Boyd) and by those who attack it (such as Laudan and van Fraassen). This will turn out to be important. But we have yet to get clear what the Ultimate Argument actually is.

Popper said that scientific realism could explain science's novel predictive successes while instrumentalism could not. Putnam, warming to the idea that realism *explains* things, says that it is "an over-arching scientific hypothesis".[12] This is odd. A philosophical view about science is to explain historical facts about science. Realism, as presented so far, is the view that science aims at true theories, that sometimes it is reasonable tentatively to presume that this aim has been achieved, and that the best reason we have to presume this is novel predictive success. Thus characterised, realism explains nothing about the history of science. In particular, realism does not explain why some scientific theories have had novel predictive success.

Perhaps what does the explaining is not the philosophical generalities of scientific realism at all. Perhaps what does the explaining are specific realist conjectures that some scientific theory is true (or nearly so). Perhaps what we have (in the simplest case) are explanations of the following kind:

> Theory T is true.
> Theory T yielded several novel predictions.
> Therefore, T's novel predictions were also true.

Is this an explanation? Well, its (alleged) *explanandum* certainly follows from its (alleged) *explanans*, as we require. *Pace* Laudan, no fallacy of affirming the consequent is involved. And as in all non-circular explanations, its (alleged) *explanans* transcends its (alleged) *explanandum*. Should the realist proffer an (alleged) *explanans* of this kind? As characterised above, the realist thinks that novel predictive success is the best reason tentatively to presume truth. And now what is claimed is that the presumed truth of theory *explains* novel predictive success. (Laudan calls these two the realist's 'upward' and 'downward' paths [13].)

Putnam formulates the realist explanation of science's success roughly as I have done:

> . . . the typical realist argument against idealism is that it makes the success of science a *miracle*. Berkeley needed God just to account for the success of beliefs about tables and chairs (and trees in the Quad) . . . And the modern positivist has to leave it without explanation (the realist charges) that "electron calculi" and "space-time calculi" and "DNA calculi" correctly predict observable phenomena if, in reality, there are no electrons, no curved space-time, and no DNA molecules. If there are such things, then a natural explanation of the success of these theories is that they are *partially true accounts* of how they behave. . . . But if these objects don't really exist at all, then . . . it is a *miracle* that a theory which speaks of curved space-time successfully predicts phenomena . . .[14]

Here Putnam appeals to partial truth instead of truth, a complication I shall ignore for the moment. He does not emphasise novel predictive success as one should, a complication I shall also ignore. Further, Putnam directs the argument against the Berkeleyan idealist and the positivist, both of whom assert the strong negative thesis that the 'theoretical entities' postulated in science *do not exist*. As a result, Putnam's main point here is that electrons, curved space-time, and DNA molecules *do exist* and that this explains why theories about them are successful.

Yet it is important to see that it is not the mere existence of 'theoretical entities', not the mere fact that 'theoretical terms' have referents, which can explain success. Laudan notes Putnam's emphasis on reference for theoretical terms and attributes the following explanation to him:

> A theory whose central terms genuinely refer will be a successful theory.
> All the central terms in theories in the mature sciences do refer.
> Therefore, the theories in the advanced or mature sciences are successful.[15]

The argument is vague: can we locate the *central* terms of a theory in a non-circular way?; can we locate the *mature* sciences in a non-circular way? Laudan does not dwell on this. He accepts that the argument is valid and even that its conclusion is true. But he thinks it a poor explanation because its premises, especially the first, are obviously false. Laudan seeks to show this on both historical and philosophical grounds.

On the historical side, he gives examples of theories whose 'central terms' referred (or so we now think) but which were unsuccessful: chemical atomism in the 18th century, Prout's hypothesis in the 19th century, the theory of continental drift in the early 20th century. (It would be better to speak of '18th century chemical atomism', and so on, to make it clear that we are speaking of *different* theories than later successful ones such as 19th century chemical atomism.) Laudan also gives examples of theories whose 'central terms' did not refer (or so we now think) but which were successful: Ptolemaic astronomy, phlogiston theory, 19th century ether theories.

Now one might quarrel with Laudan's claims about some of these historical examples. How successful was phlogiston theory, for example? Such quarrels would be intensified if emphasis was placed on *novel* predictive success (though neither Putnam nor Laudan give any emphasis to this). How much *novel* predictive success did Ptolemaic astronomy have, for example?

But we need not pursue any of these historical questions. For Laudan has a simple and devastating philosophical argument which divorces successful reference from success. We can easily construct a referring theory which will be unsuccessful: take a successful referring theory, retain its existential claims, and negate its theoretical ones. "Richard Nixon is tall, blonde, honest and never swears" refers to Richard Nixon all right, but it says a lot of false things about him and would be very unsuccessful in predicting Nixon-phenomena. Obviously, in any realist explanation of science's success it is truth or near-truth which is going to be important, rather than mere successful reference.

This does not mean that all the ink spilled over the reference of theoretical terms has been wasted ink. Realists think that theories typically assert the existence of their 'theoretical entities', so that successful reference is typically a *necessary condition* for truth or near-truth. However, it is equally important for realists that reference, while a necessary condition for truth, is not a *sufficient* condition. Realists hold that we know more about, say, electrons than our ancestors did, that while our ancestors had false theories about electrons we have true (or truer) ones. But we can only say this if the false theories of our ancestors referred to electrons just as our own theories do. A theory may be referential yet false.

But if reference is only a necessary but not a sufficient condition for truth, then it is clear that it is hopeless to try to explain why a theory is successful merely by pointing out that its theoretical terms do successfully refer. If the realist is to explain the success of some scientific theory, it is truth (or near truth) that is needed rather than mere successful reference. Putnam's explanation is of this kind: it is because electrons exist *and* electron-theory gives a true (or partially true) account of them that electron-theory is successful. And the argument is that the Berkeleyan or positivist, who denies the existence of electrons, can give no account of the success of electron-theory.

An immediate worry about the argument is that Putnam has chosen his opponents carefully. Not every anti-realist is a Berkeleyan or a positivist (in Putnam's sense). There are anti-realists who do not *deny* the existence of 'theoretical entities', but who prefer to remain *agnostic* on the matter and fashion their philosophy accordingly. It remains to be seen whether Putnam's argument can be directed against them too.

Before we turn to that question, we can at last be clear about what the Ultimate Argument actually is. It is an example of a so-called *inference to the best explanation*. How, in general, do such inferences work?

The intellectual ancestor of inference to the best explanation is Peirce's *abduction*. Abduction goes something like this:

> *F* is a surprising fact.
> If *T* were true, *F* would be a matter of course.
> Hence, *T* is true.

The argument is patently invalid: it is the fallacy of affirming the consequent again. One might say (nobody has) that although abduction is *deductively* fallacious, it is actually a perfectly valid argument in a special abductive or ampliative or inductive logic. But it conduces to clarity if we say instead that abduction is a deductive *enthymeme* and supply its missing premise. Its missing premise is obviously the (metaphysical) principle that any explanation of a surprising fact is true. This conduces to clarity because we can now see clearly that abduction is something no sane philosopher should accept. The metaphysical premise which validates the inference is obviously *false*. Any sane philosopher knows of countless cases where an explanation of some surprising fact is false.[16]

But what if an explanation of some surprising fact is *better* than any other explanation that we have? Inference to the best explanation goes something like this:

> *F* is a fact.
> Hypothesis *H* explains *F*.
> No available competing hypothesis explains *F* as well as *H* does.
> Therefore, *H* is true.[17]

This argument too is patently invalid. Most say that although inference to the best explanation is *deductively* invalid, it is actually a perfectly valid argument in a special abductive or ampliative or inductive logic. But again it conduces to clarity if we say instead that inference to the best explanation is a deductive *enthymeme* and supply its missing premise. Its missing premise is obviously the (metaphysical) principle that the best available explanation of any fact is true. This conduces to clarity because we can now see clearly that inference to the best explanation thus construed is something that no sane philosopher should accept. Again, the metaphysical premise which validates the inference is obviously *false*. Any sane philosopher knows of countless cases where the best available explanation of some fact turned out to be false.

Reconstructing inference to the best explanation as a deductive *enthymeme* conduces to clarity in another way — it gives us a clue as to how the inference might be rescued from absurdity. It is absurd to say that the best available explanation of any fact is true. It is not obviously absurd to say that it is reasonable to accept the best available explanation of any fact as true (tentatively, of course), or to presume (tentatively) that it is true. For it is plain, is it not, that it may be reasonable tentatively to accept something as true which subsequently turns out to be false. (If it does turn out to be false, we say that what we accepted was wrong, not that we were wrong to accept it.) This suggests that we replace the obviously false metaphysical premise of the argument by this epistemological premise (amending the conclusion accordingly). The resulting pattern of argument is deductively valid and its major premise is not obviously mistaken.

Inference to the best explanation, thus reformulated, will still not quite do. What if our best explanation of some fact is a perfectly lousy one? Would it be reasonable to accept it tentatively as true? Obviously not. What we need is a principle to the effect that it is reasonable to accept a *satisfactory* explanation which is the best we have as true. And we need to amend the inference-scheme accordingly. What we finish up with goes like this:

> It is reasonable to accept a *satisfactory* explanation of any fact, which is also the *best* available explanation of that fact, as true.
> *F* is a fact.
> Hypothesis *H* explains *F*.
> Hypothesis *H satisfactorily* explains *F*.
> No available competing hypothesis explains *F* as well as *H* does.
> Therefore, it is reasonable to accept *H* as true.

Of course, for this argument-scheme to be applicable, the 'explanationist' owes us an account of when an explanation is minimally adequate (or 'satisfactory'), as well as an account of when one satisfactory explanation is *better* than another. But this digression on inference to the best explanation has gone on long enough, so I will simply assume that such explanationist accounts can be given.

To return to the Ultimate Argument for scientific realism. It is, I suggest, an inference to the best explanation. The fact to be explained is the (novel) predictive success of science. And the claim is that realism (more precisely, the conjecture that the realist aim for science has actually been achieved) *explains* this fact, explains it *satisfactorily*, and explains it *better* than any non-realist philosophy of science. And the conclusion is that it is reasonable to accept scientific realism (more precisely, the conjecture that the realist aim for science has actually been achieved) as true.

Suppose that we now have the argument right. If so, to repeat, it is not realism that explains facts about science, and realism is not an "over-arching scientific hypothesis". If realism could explain facts about science, then it could be refuted by them too. But a philosophy of science is not a description or explanation of facts about science. It is fashionable to identify scientific realism with the view that all (or most) scientific theories are true (or nearly so), or with the view that all (or most) *current* scientific theories are true (or nearly so), or with the view that all (or most) *current* theories in the '*mature*' sciences are true (or nearly so). But a pessimistic scientific realist might think none of these things *without thereby ceasing to be a realist.* A slightly more optimistic realist might tentatively accept some particular theory as true. And the suggestion is that such a realist can then give the best explanation of that particular theory's success.

Is the suggestion correct? That partly depends upon whether it is true that non-realists have no explanation, or only an inferior explanation, of (novel) predictive success. As we have seen, Putnam directed his argument against Berkeley and the 'positivist'. It seems to be right that Putnam's positivist (for whom no theoretical entities exist and for whom all theories are false) can only regard (novel) predictive success as a lucky accident or 'miracle'. We think poorly of a person who 'explains' why the light goes on when we press the switch by saying "It is just a lucky accident". And we should think equally poorly of the positivist who says the same thing of science's (novel) predictive success.

The case of Berkeley is more interesting. Berkeley denies not only the theoretical entities of science, but also the 'theoretical entities' of commonsense realism, tables and chairs and trees in the Quad. (He tries to soften the latter denial by re-defining words like 'table' and by telling us a tale about what we 'really mean' by such statements as "My table is in my study though nobody is perceiving it". No matter.) But if there are no tables and chairs and trees in the Quad, how come our false beliefs about such things are so successful? Such commonsense beliefs yield innumerable successful predictions: "If I return to my study, I shall again see my table and chair", "If you come into the Quad with me, we shall both see the tree", "If I shut my eyes for a second, when I re-open them I will see things as I do now", and so on. As Putnam notes, Berkeley gives a *theological* explanation: God directly causes our perceptions, God is good, so God causes our perceptions in a regular fashion. Berkeley would deny that the success of commonsense realist beliefs is a miracle: phenomenal regularities are only to be expected, given Berkeley's metaphysic. What would be miraculous would be a 'sensible thing' which looked, smelled and felt like an orange, but tasted like a banana. God might work such a miracle. But so as not to confuse us, He does not (or not often).

Berkeley's theoretical posit (God) introduces all sorts of problems which the commonsense realist's posits (independently existing objects) do not. Hence it is widely (and rightly) regarded as the weak link of his system. But if you remove God from Berkeley's picture, you have a metaphysic (phenomenalism) which has no explanation of the success of commonsense beliefs. If Berkeley's theory and phenomenalism were the only available theories, an inference to the best explanation should lead us to prefer Berkeley!

As for science, Berkeley takes a thoroughgoing instrumentalist view of it. (In my opinion, he was the first to do so, *pace* Duhem and his countless followers.[18]) So what can Berkeley make of the predictive successes of science? He can say that it is no accident that 'mathematical hypotheses' contrived to summarise some phenomenal regularities should successfully predict new instances of those regularities. That just testifies to man's ingenuity (in the contriving) and to God's benevolence (in the maintaining of the known phenomenal regularities). We cleverly concoct a fiction called 'geometrical optics', which trafficks in non-existent light rays, and in 'mathematical hypotheses' concerning the rectilinear propagation, reflection and refraction of these non-entities, to summarise phenomenal regularities about things casting shadows, how things look in mirrors, sticks looking bent in water, and so on. The regularities being correct (God willing), and the fiction being cooked up to yield them, it is no accident that it successfully predicts future instances of them.

But what if geometrical optics yields a *new* regularity? What if it predicts that looking at a thing through a certain arrangement of lenses will make it look bigger? (I know that the example lacks historical veracity.) The realist who accepts geometrical optics as true will expect this prediction to be true also. Berkeley can have no such expectation. For all he knows, God could fix it that objects viewed through telescopes will look smaller, disappear altogether, turn into ducks, or whatever. Only after Berkeley has learned from experience that they do none of these things, but look bigger instead, can he say "Ah, that is how God's benevolence manifests itself in this case". But he could not have predicted it — and he could not have explained it, in terms of the truth of geometrical optics, either. (It may be objected that Berkeley could explain the novel predictive success of geometrical optics in terms of its *empirical adequacy*. I am not aware that Berkeley did or could give such an explanation. I consider it soon.)

So I think that Putnam is right. The realist can give a better explanation of science's (novel) predictive success than either the positivist or the Berkeleyan idealist. (This is *not* to say that the realist's explanation is a good one.) But what of other anti-realists, such as van Fraassen or Laudan? They do not deny (as the instrumentalist does) that theories are either true or false. They do not assert (as the positivist does) that they are all false — they concede that some theories might be true. What they deny is that it can ever be reasonable to presume (however tentatively) that any theory is true. Accordingly, they do not think true theories are a sensible aim for science, and they put something else in its place. They are anti-realists on epistemological grounds; we might call them *epistemological anti-realists.* What explanation might they give of (novel) predictive success?

Van Fraassen replaces truth by empirical adequacy as an aim for science. A theory is empirically adequate if all of its 'observational consequences' are true. So an explanation which van Fraassen might give and which parallels the realist explanation is:

> Theory *T* is empirically adequate.
> Theory *T* yielded several novel predictions.
> Therefore, *T*'s novel predictions were true.

This 'explanation' invokes the fact that all of a theory's observational content is true to explain why some particular observational consequences are. This is like explaining why some crows are black by saying that they all are. The realist explanation seems better than this, because the postulated truth of a theory (and the implied existence of its theoretical entities) transcends the truth of some or all of its observational consequences. (One wonders how the empirical adequacy of a theory might be explained if not by postulating its truth.)

In fact, van Fraassen offers us a quite different explanation of science's predictive success. The success of current scientific theories is no miracle, he says, because only successful theories survive the fierce Darwinian competition to which all theories are subjected.[19] But this changes the subject. It is one thing to explain why only successful theories survive, and quite another thing to explain why some particular theory is successful. van Fraassen's Darwinian explanation of the former can be accepted by realist and anti-realist alike.[20] But it yields no explanation at all of the latter. You do not explain why (say) electron-theory is (scientifically) successful by saying that if it had not been it would have been eliminated. Just as you do not explain why (say) the mouse is (biologically) successful by saying that if it had not been it would have been eliminated. Biologists explain why the mouse is successful by telling a long story about its well-adaptedness. Realists want to explain why electron-theory is successful by telling a shorter story about its 'well-adaptedness', that is, its truth.

Laudan replaces truth by problem-solving ability as an aim for science. A theory solves an empirical problem if it yields a correct answer to it. So Laudan might give an explanation of success like the following:

> Theory *T* correctly solves all its empirical problems.
> Theory *T* yields several novel predictions.
> Therefore, *T*'s novel predictions are true.

Again, this 'explanation' invokes the fact that all of a theory's empirical consequences are true to explain why some particular ones are. Again, this is like explaining why some crow is black by saying that they all are. Again, the realist explanation seems better than this. And again, one wonders how the problem-solving ability of a theory could be explained without postulating its truth.

I should make it clear that Laudan himself does not propose an explanation of this kind, or indeed of any other kind. He says that the realist explanation is "attractive because self-evident". But he objects to it on epistemological grounds, saying that we can never "reasonably presume of any given scientific theory that it is true".[21] Further, he argues historically that past theories which we now think false (and non-referential) were just as successful as present theories which realists think true (and referential). Given such views, Laudan must think that success just is a lucky accident and eschew all attempts to explain it.

Finally, let us consider what Jarrett Leplin calls the *surrealist explanation* of success, 'surrealism' being short for 'surrogate realism'. It goes like this:

> The world is *as if* theory *T* were true.
> Theory *T* yields several novel predictions.
> Therefore, *T*'s novel predictions are true.

Is this explanation as good as the realist one?

It is not easy to answer this question, because it is not easy to say what "The world is *as if* theory *T* were true" actually *asserts*. For the explanation to go through it must assert *at least* that the world is *observationally* as if *T* were true. If it asserts no more than this, then it is just a fancy way of saying that *T* is observationally or empirically adequate. In this case, the surrealist explanation comes from the same stable as those already considered, and is subject to the same objections.

So perhaps "The world is *as if* *T* were true" is meant to be more than a fancy way of saying "*T* is empirically adequate". Perhaps it is meant to entail everything that *T* entails *except* just for *T* itself. But this is not a coherent position. Let *S* be any statement entailed by but not entailing *T*. On the view suggested, "The world is *as if T* were true" entails both *S* and "Either *T* or not-*S*". But these in turn entail *T* (by double negation and disjuctive syllogism). Hence "The world is *as if T* were true" entails *T* also (by transitivity of entailment). Given the logical principles just mentioned, "The world is *as if T* were true" cannot entail everything that *T* entails except just for *T* itself.

The surrealist's dilemma is plain. If he invokes less than empirical adequacy, then he has no explanation of empirical success.[22] If he invokes empirical adequacy and no more, then he has only a poor explanation. If he invokes more than empirical adequacy, then he has to tell us *what* more in a way that does not collapse "The world is *as if T* were true" into "*T* is true".

It seems, then, that the realist has a better explanation of novel predictive success than the epistemological anti-realists. Van Fraassen's empirical adequacy and Laudan's problem-solving ability and the surrealist's *as if* ploy all yield alternative descriptions of empirical success in general. As such, they do not give good explanations of particular instances of it. Positivistic *atheism* about theories and their theoretical entities makes a mystery of the novel predictive success of those theories. And so does the *agnosticism* about theories and their theoretical entities recommended by epistemological anti-realists. The realist explanation seems better because it posits the truth of a successful theory and the existence of its theoretical entities.

In any case, realists will be thoroughly impatient with the theoretical agnosticism of these anti-realists and with the rival 'explanations' to which it leads. Those 'explanations' are clearly parasitic upon the straightforward realist explanation (as emerges most clearly in the surrealist case). Moreover, impatience can prompt argument here. Anti-realisms, atheistic or agnostic, must all operate with some distinction between observation and theory, or between 'observable entities' and 'theoretical entities'. Without that distinction, truth cannot be distinguished from empirical adequacy or from problem-solving ability, and surrealism collapses into realism. Anti-realists draw the observable/theoretical line in different places — but they all draw it somewhere *and* give it crucial ontological and/or epistemological significance. Now realists steadfastly argue that no such distinction can be drawn, at least none sharp enough to bear the ontological and/or epistemic burdens which anti-realists place upon it. The distinction between what we happen to be able to observe and what not is irredeemably vague. And why should my ontological commitments be limited to the 'observable' or my epistemic commitments to statements about the 'observable'? This is not the place to rehearse these familiar realist arguments.[23] Suffice it to say that anyone persuaded by them will object that the explanations we have pitted against the realist explanation all rest upon dubious and human-chauvinistic philosophy.

I concluded a paragraph back that the realist explanation of success *seems* better than anti-realist ones because it posits truth and reference for successful theories. But is it really any better? We objected to explaining why *some* empirical consequences of a theory are true by invoking the fact that they *all* are. Cannot a similar objection be levelled at the realist? The realist explanation is "Theory T is true", which is the same as saying "All the consequences of T are true". So the realist explains why some consequences are true by saying that they all are. The only difference between the realist and anti-realist explanations is one of *scope*: the realist invokes the fact that all of a theory's consequences are true, the anti-realist invokes the fact that all of its empirical consequences are true.

The realist explanation is better, one might say, because broadness of scope is an explanatory virtue. Other things being equal, we prefer the broader explanatory theory because it tells us more, excludes more possible states of affairs, is more testable. Whatever the virtues of this maxim in science, its application to our case is problematic because our case is a curious one. The realist explanation in terms of truth is *not* more testable than the anti-realist explanation in terms of empirical adequacy. The realist explanation may 'tell us more', but in the nature of the case there can be no *empirical evidence* that the more it tells us is correct.

Such reflections convince Arthur Fine that the realist explanation is actually *worse* than the anti-realist one. He writes:

Metatheorem I. If the phenomena to be explained are not realist-laden, then to every good realist explanation there corresponds a better instrumentalist one.

Proof: In the proffered realist explanation, replace the realist conception of truth by the pragmatic conception [of truth as empirical adequacy]. The result . . . will be the better instrumentalist explanation.[24]

Fine's intuition (and here he follows van Fraassen) is that the realist's explanation of success involves 'metaphysical excess baggage', since there can be no *evidence* for truth over and above empirical adequacy. In science we try to avoid encumbering theory with ingredients which demonstrably have no empirical pay-off — why not do the same in *philosophy* of science?

In reply to this, the realist can simply say that there are explanatory virtues which are neither evidential nor obviously connected with scope. In science and in philosophy of science, one of a pair of empirically equivalent theories can possess explanatory virtues that the other lacks. Ancient astronomers thought that the stars move as they do because they are fixed on a sphere which rotates once a day around the central earth. Compare that theory with its surrealist transform, the theory that stars move *as if* they were fixed on such a sphere. Is not the first theory explanatory in a way that the second is not, despite the fact that the second is expressly designed to be empirically equivalent with the first? Nineteenth-century geologists devised an elaborate theory of fossil formation, to which twentieth-century geologists have added an equally elaborate theory of radio-carbon dating. Call the amalgam of these theories G, and compare it with its Gossian transform G^*, the theory that God created the universe in 4004 BC in such a way that it would *appear* that G was true. Is not G explanatory in a way that G^* is not, despite the fact that G^* is expressly designed to be empirically equivalent with G? And has not G^* been rejected in favour of G, despite the fact that no empirical evidence can decide between them?[25] We are actually in old logical positivist territory here (as Wade Savage has pointed out to me): compare *any* scientific theory T with its Craigian transform T_C; is not T explanatory in a way that T_C is not, despite the fact that T_C is expressly designed to be empirically equivalent with T? We are just re-running Hempel's 'theoretician's dilemma' all over again, and the battle-lines are the same as they always were. If the name of the game is 'saving the phenomena' (van Fraassen, Fine), then one of a pair of empirically equivalent theories is just as good as the other. If the name of the game is explaining the phenomena (realism), then this is not the case.

As in science, so also in philosophy of science. Compare the realist explanation of science's success in terms of truth and reference, with the anti-realist explanation in terms of empirical adequacy. Is not the former explanatory in a way that the second is not, despite the fact that the second was expressly designed to be empirically equivalent with the first?

The thrust of this realist rhetoric is the same both at the scientific and at the meta-scientific levels. It is that explanatory virtues need not be evidential virtues. It is that you should feel *cheated* by "The world is *as if T* were true", in the same way as you should feel *cheated* by "The stars move *as if* they were fixed on a rotating sphere". Realists do feel cheated in both cases. But anti-realists do not. If you are an anti-realist who does not mind surrealist and other transforms of *scientific* theories, then you will not be impressed by the Ultimate Argument on the meta-scientific level either.

Michael Levin has a deeper worry about the Ultimate Argument, indeed, about the entire project of giving a meta-scientific explanation of science's success. He claims that *science* can explain its own success, and that we do not need philosophy of science (whether realist or anti-realist) to do this:

> The explanation of the success of a theory lies within the theory itself. The theory itself explains why it is successful . . .
>
> A theory's successes are the true predictions it has made with its own internal resources. Conjoin them and you have its success, but you do not have any further phenomenon which the theory in question fails to explain and which may perhaps be explained by some such other hypothesis as truth. To explain a conjunction, explain its conjuncts.[26]

But this cannot be *quite* right. Granted, a theory itself explains its predictive successes (assuming it is not a surrealist theory). But what needs explaining here are not those predictive successes, facts about the world, but rather the fact that the theory had those successes, a fact about the theory. A theory itself cannot "explain why it is successful": electron-theory, for example, is about electrons, not about electron-theory.

It transpires that Levin's real worry is whether a theory's *being true* could explain anything about that theory. He argues that scientific theories are (intellectual) artefacts, that an explanation of the success of an artefact is always *mechanistic*, and that truth is not a mechanism:

> And here is my problem: what kind of *mechanism* is truth? How does the truth of a theory bring about, cause or create, its issuance of successful predictions? Here, I think, we are stumped. Truth . . . has nothing to do with it. "By being true" never satisfactorily answers the question, Why did such and such a belief lead to correct expectations? The answer always lies elsewhere.[27]

Is it true that the truth of a belief never explains why that belief led to correct expectations? Suppose Hopalong succeeds in finding gold in them-thar-hills. How might we explain his success? A natural explanation (though not the only one) is that Hopalong believed that there was gold in them-thar-hills, acted accordingly, *and that his belief was true.* But semantic descent being what it is, we might as well say that Hopalong believed that there was gold in them-thar-hills, acted accordingly, *and that there was gold in them-thar-hills.* Thus Levin: it was not the truth of Hopalong's belief that made it successful, rather it was the fact that there was gold in them-thar-hills just as Hopalong believed there to be.

This is playing with words. Semantic *ascent* being what it is, we do not have *rival* explanations here, but rather equivalent formulations of the *same* explanation. "*H* believed that *G* and *G*" is equivalent to "*H* believed *truly* that *G*" (given the theory of truth that Levin and the realists both accept).

Levin insists upon the first formulation because of his worries about truth not being a 'mechanism'. Of course truth is not a mechanism. It is a property which beliefs or theories may possess. But when a belief possesses it, this fact can figure in an explanation of why acting upon that belief leads to success. Such an explanation can even be described as 'causal' or 'mechanical'. Levin insists that "By being true" cannot explain why Hopalong's belief was successful, that the answer "lies elsewhere", presumably in them-thar-hills. But "By being true" takes us to them-thar-hills and tells us that gold lies there.

So I do not insist, as Levin does, on semantic descent. It is worth noting that if we *do* insist upon it, we will be equally sceptical of the anti-realist explanations of success that we have considered. If truth is not explanatory because it is not a 'mechanism', then neither, presumably, is empirical adequacy. More important, suppose we follow Levin and say that the *real* explanation of the success of electron-theory (say) is that there really are electrons, they really do carry a certain elementary charge, and so on. This explanation yields everything that the realist wants to say about electron-theory (say). It is just that Levin forbids him from ascending to *say* it. He insists upon "Electrons really exist" rather than "The term 'electron' really refers", and upon "Electrons carry a certain elementary charge" rather than "The statement 'Electrons carry a certain elementary charge' is true". The realist, puzzled, might easily comply.

So what, ultimately, of the Ultimate Argument? It is best construed as an inference to the best explanation of facts about science. The facts which need explaining are best construed as facts about the *novel* predictive success of *particular* scientific theories. The realist explanations of such facts are best construed as invoking (conjecturally) the truth of those theories (or their near-truth if we can develop such a notion) and the reference of their theoretical terms. Positivistic anti-realists have no competing explanation of such facts about science. Epistemological anti-realists give no competing explanation either. But we can give such explanations on their behalf. And when we do, we find that the situation is curiously circular: realist explanations of success are preferable *on realist grounds*; anti-realist explanations of success are preferable *on anti-realist grounds*. The attempt to make realism explanatory of facts about science, which is what the Ultimate Argument does, will fail to convince anti-realists who doubt that science itself is explanatory.

University of Otago

NOTES

[1] It was christened thus by van Fraassen in his (1980), p. 39.
[2] Paraphrased from Putnam, (1975), p. 73.
[3] As cited by Blake (1960), p. 34 (the passage comes from Clavius's *Commentary on the Sphere of Sacrobosco*).
[4] As cited by Blake (1960), p. 33.
[5] Attempts to rebut this sceptical argument by Clavius and Kepler are documented in Jardine's (1979).
[6] Laudan (1981), p. 45.
[7] Laudan's accusation here is not without foundation. Brown explicitly reconstructs the Ultimate Argument (which he calls the 'miracle argument') as an argument *for* the (probable) truth of theories which make true predictions: see his (1982), pp. 98—9, or his (1985), p. 51. The argument Brown reconstructs is deductively invalid and is a souped-up version of affirming the consequent.

But no fallacy *need* be involved in taking predictive success as a reason for believing (tentatively) in the truth of a theory. The argument can be reconstructed as follows:

> If a theory is predictively successful, then it is reasonable to accept it tentatively as true.
> Theory T is predictively successful.
> Therefore, it is reasonable to accept theory T tentatively as true.

This argument is *logically* impeccable. Sceptical doubts about it must focus upon the 'inductive principle' which forms its major premise. But you do not refute that premise by pointing out that predictive success does not entail truth: it may be reasonable tentatively to accept a theory as true even though that theory subsequently turns out to be false. I mention this because Laudan, in criticising realism on sceptical grounds in his (1981), thinks it sufficient to dispose of the idea that predictive success *entails* truth. (I do not myself think that predictive success *simpliciter* is a reason for acceptance, and I will be saying why.)

Here, as elsewhere, I prefer to construe so-called 'inductive arguments' as deductive arguments with 'inductive principles' of one kind or another among their premises. This conduces to clarity and obviates the need for any special inductive *logic*.

[8] Whewell (1837), volume II, p. 68.

[9] For an elaboration and defence of it, see Worrall's (1985).

[10] Popper (1963), pp. 117—8. Popper obviously has *successful* novel predictions in mind here, for an *un*successful one would hardly count as a discovery.

[11] Duhem (1954), p. 28; see also pp. 297ff.

[12] Putnam (1978), p. 19.

[13] Laudan (1981).

[14] Putnam (1978), pp. 18—19.

[15] Laudan (1981), p. 23.

[16] Peirce did not formulate abduction as I have. The only important difference is that Peirce's conclusion is not "*T* is true" but rather "There is reason to suspect that *T* is true" [Peirce (1931—58), 5.189]. Peirce's original scheme is also invalid, and the missing premise that would validate it is also unacceptable. Is there reason to suspect that any explanation, *however bizarre*, of some suprising fact is true? However, Peirce's intuition that abductive arguments are *epistemological* (as evidenced by the epistemic modifier in his conclusion) was sound: I shall be saying the same of inference to the best explanation. (Incidentally, both abduction and the misleadingly labelled 'inference *to* the best explanation' are located firmly in the context of justification rather than the context of discovery, despite what many think.)

[17] This is a slightly simplified version of Lycan's formulation in his (1985), p. 138. Some make truth a defining condition of explanation, so that we do not have an explanation *at all* unless what does the explaining is true. They would have to reformulate the argument so that it becomes inference to the best *putative* explanation. I prefer to make truth an adequacy condition on explanation rather than a defining condition of it, so that it makes sense to speak of a false explanation.

[18] The orthodox view is that an instrumentalist tradition regarding astronomical hypotheses was inaugurated by the great Hellenic astronomers (Eudoxus, Hipparchus, Apollonius, and Ptolemy). I criticise that orthodoxy in my (1981).

[19] van Fraassen (1980), pp. 39—40. For a further discussion of this explanation, and of the biological analogy on which it is based, see my (1985), pp. 209—10.

[20] It was actually proposed by the realist Popper in 1934: see Zahar (1983), p. 169. Incidentally, Brown says [in his (1985), p. 49] "Karl Popper has steadfastly held that the success of science is not to be explained; it is a miracle". But what Popper holds is that no theory of knowledge should try to explain *how we have come up with* successful theories sometimes. This is quite consistent with explaining why a particular theory is successful, and with explaining why only successful theories survive. In this area it is crucial that we get the *explanandum* right.

[21] Laudan (1981), p. 30. Laudan does not actually argue for this strong epistemic

thesis: rather, he seems to think that it follows from a weaker thesis that he does argue for, the thesis that the evidence does not *entail* that a scientific theory is true. But the latter thesis does not entail the former at all, of course (see also footnote 7 above). Indeed, it can be shown that Laudan himself thinks that we can reasonably presume of *some* theories that they are true: see my (1979), pp. 459—60.

[22] Actually, the surrealist *as if* ploy is sometimes used to invoke less than empirical adequacy. Historians of astronomy say that Eudoxus cannot have thought his theory of planetary motion true, because 'interpreted realistically' it clashes with the observed fact that the planets vary in brightness. Eudoxus was saying 'The world is *as if E*' rather than just *E* (where *E* is Eudoxus's theory). But if 'The world is *as if E*' is not to clash with observed brightness variations, it cannot mean 'The world is *observationally as if E*' (or '*E* is empirically adequate'). It must mean something like 'As far as planetary positions go (but not their brightnesses), the world is *as if E*'. Here the surrealist ploy eliminates some of the empirical content of the theory to which it is applied. A similar case arises if it is applied (as it has been by some historians) to Ptolemy's theory of the moon.

[23] I rehearse some of them against van Fraassen in my (1985), especially pp. 204-9, and against Laudan in my (1979).

[24] Fine (1986), p. 154. According to Fine, anti-realists like van Fraassen or Laudan do not so much *replace* truth by empirical adequacy as an aim for science, as retain truth as the aim but give an *empirical adequacy theory of truth.* One can see them this way. It is not the way they see themselves, nor is it the clearest way to see them. But this issue does not affect the matters being discussed here, so I shall not pursue it.

[25] It is not that *G** yields no explanation *at all* of fossils and decay elements in rocks, it is merely that it yields a quite *different* explanation than *G* does. *G**'s explanatory mechanism is *essentially* divine and unintelligible by humans, while for its explanatory *details* it is entirely parasitic upon *G* (converting them into details about what God had in mind on Monday — or was it Tuesday? — one week 4004 years ago). Science rejects *G** in favour of *G*, despite their empirical equivalence. Perhaps this is because *G**'s essential mechanism is (and is meant to be) unintelligible. Perhaps it is because the empirical character of *G** is a sham: if future geologists should abandon *G* in favour of *H*, the Gossian will happily switch to *H** and preserve the *essence* of his position. Perhaps science prefers *G* to *G** for a mixture of these two reasons, since they are not unconnected reasons: it is because it is unintelligible that *G**'s essential mechanism can be preserved. The important point for my purposes is only that we do here have a rational choice between empirically equivalent theories. (It may be objected that Gosse's *original* hypothesis, unlike *G**, was not empirically equivalent with nineteenth century theories of fossil formation. Perhaps. Still, was it rejected on *empirical* grounds?)

[26] Levin (1984), p. 127.

[27] Levin (1984), p. 126.

REFERENCES

Blake, R. M. (1960) 'Theory of Hypothesis Among Renaissance Astronomers', in Blake, R. M. (ed.), *Theories of Scientific Method*, Seattle and London: University of Washington Press, pp. 22—49.

Brown, J. R. (1982) 'Realism, Miracles, and the Common Cause', in Asquith, P. D. and Nickles, T. (eds.), *PSA 1982 Volume I*, Michigan: East Lansing, Philosophy of Science Association, pp. 98—106.

Brown, J. R. (1985) 'Explaining the Success of Science', *Ratio* **27**, 49—66.

Duhem, P. (1954) *The Aim and Structure of Physical Theory*, Princeton: Princeton University Press.

Jardine, N. (1979) 'The forging of Modern Realism: Clavius and Kepler Against the Sceptics', *Studies in History and Philosophy of Science* **10**, 141—173.

Laudan, L. (1981) 'A Confutation of Convergent Realism', *Philosophy of Science* **48**, 19—49.

Levin, M. (1984) 'What Kind of Explanation Is Truth?', in J. Leplin (ed.), *Scientific Realism*, Berkeley and Los Angeles: University of California Press, pp. 124—139.

Lycan, W. G. (1985) 'Epistemic Value', *Synthese* **64**, 137—164.

Musgrave, A. E. (1979) 'Problems with Progress', *Synthese* **42**, 443—464.

Musgrave, A. E. (1981) 'Der Mythos vom Instrumentalismus in der Astronomie', in H-P. Duerr (ed.), *Versuchungen: Aufsätze zur Philosophie Paul Feyerabends*, Frankfurt: Suhrkamp Verlag, 2 Band, pp. 231—279.

Musgrave, A. E. (1985) 'Realism Versus Constructive Empiricism', in P. M. Churchland and C. A. Hooker (eds.), Chicago and London: *Images of Science*, University of Chicago Press, pp. 197—221.

Peirce, C. S. (1931—58) *The Collected Papers of Charles Sanders Peirce*, ed. C. Hartshorne and P. Weiss, Cambridge, Mass.: Harvard University Press.

Popper, K. R. (1963) *Conjectures and Refutations*, London: Routledge and Kegan Paul.

Putnam, H. (1975) *Mathematics, Matter and Method* (*Philosophical Papers, Volume I*), London: Cambridge University Press.

Putnam, H. (1978) *Meaning and the Moral Sciences*, London: Routledge and Kegan Paul.

van Fraassen, B. C. (1980) *The Scientific Image*, Oxford: Clarendon Press.

Whewell, W. (1837) *History of the Inductive Sciences*, London: John W. Parker.

Worrall, J. (1985) 'Scientific Discovery and Theory-confirmation', in J. C. Pitt (ed.), *Change and Progress in Modern Science*, Dordrecht: D. Reidel, pp. 301—331.

Zahar, E. (1983) 'The Popper-Lakatos Controversy in the Light of "Die Beiden Grundprobleme der Erkenntnistheorie"', *British Journal for the Philosophy of Science* **34**, 149—171.

RICHARD SYLVAN

RADICAL PLURALISM – AN ALTERNATIVE TO REALISM, ANTI-REALISM AND RELATIVISM

Realism, the dominant 20th century position in Anglo-American thought, is, in the relevant sense, a one world position. There exists a unique actual world, or reality, external to "us", which not only determines how things are locally and globally, but determines as well truth, and thus also uniquely fixes correctness in science, *the* correct theory being that which corresponds to reality. Anti-realisms such as idealism and phenomenalism reject, in one way or another, the tricky externality requirement. Relativism and pluralism, by contrast, reject one of the uniqueness requirements, but in significantly different ways. Relativism resists, in one fashion or another, the imposition of any ranking better than "equally good" and of any rankings warranting differential choice, on the multiple interpretations or, very differently, multiple realities or worlds disclosed. Pluralism, however, to set down at once the crucial contrast, permits and typically makes rankings, which enable choice (including realist and idealist and theist choices, among many others). Pluralism thus comes in two distinct forms: theory or meta-pluralism, according to which there are many correct theories (especially larger philosophical positions) but at most one actual world; and radical or deep pluralism which goes to the root of these differences in correctness, to be found in things, and discerns a plurality of actual worlds as well as of theories.

Realism, then, characteristically involves not only the (*existential*) claim that *there is an actual world* with various prized properties such as externality and mind-independence; but it further involves the claim that there is only one such world, that *the world is unique*. The *uniqueness* claim is essential: otherwise Reality is not fully determinate, and the actual world cannot perform expected realist functions of determining truth, correctness and the like, in a way that is single-valued and entire.[1] It is the rarely considered, but normally simply assumed, uniqueness claim that is a main focus of concern here. It will be contended that uniqueness fails, that not only is uniqueness not established, but it cannot be nonlegislatively, because there is not a unique actual world. A central thesis to emerge is then that there are many

Robert Nola (ed.), Relativism and Realism in Science, 253–291.

actual worlds, among which one may — or may not — be selected as *the* world.[2] But different choices are defensible. Thus (to cast a theme which holds for very unSchopenhauerian reasons in Schopenhauerian terms), *the* world is a manifestation of will; it involves in principle a *constrained* choice.

Anti-realism, the usual false contrast with realism[3], is accordingly not of much present concern (except insofar as radical pluralism erroneously gets accounted anti-realist!). For anti-realisms are characteristically reductionistic; they characteristically take issue, in one way or another, with the prized (transcendental) properties ascribed under the existential claim — mind or human independence especially — if not with the main claim itself. Anti-realisms will of course find a place in the pluralistic framework to be elaborated; but, especially in view of their anthropocentricism,[4] that place deserves to be a lowly and unimportant one (see further §6).

1. ON THE WORLD AND THEORY STRUCTURE OF RADICAL PLURALISM

Analytic philosophers often have a hard time understanding radical pluralism. Hopefully it will clarify matters to outline at once the doubly pluralistic framework presupposed, which is to be developed and further defended (see Diagram 1).[5]

Subscripting is adopted to indicate further multiplicity — because there are many variants upon physical realism, many styles and depths of scepticism, many cultures of varying qualities with somewhat different worlds and commonsensical theories, especially if past and future variations are duly recognised, and so on. Radical pluralism both surveys (a level up, if you like) *and* is part of this richness and comprehensible complexity. Sistological pluralism is *a* radical pluralism (a subscripted type), one based upon a liberal theory of objects. It is, in fact, an enriched commonsense. It is not the only radical pluralism, simply a preferred one, selected here (from which all superficially absolute claims are advanced). Sistological pluralism is a bundle of positions, with classification, ranking and choice of positions within the bundle, coupled with a system of worlds also classified. But it is also a position within the bundle, a chosen position where nonexistent objects have standing, which asserts that there is this bundle of positions, as already indicated.

Diagram 1: Philosophical theories and worlds as seen from, and as including, a radical pluralism.

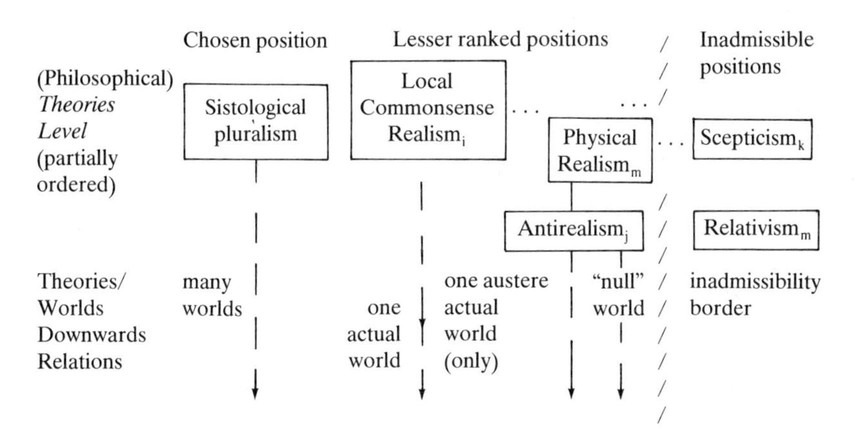

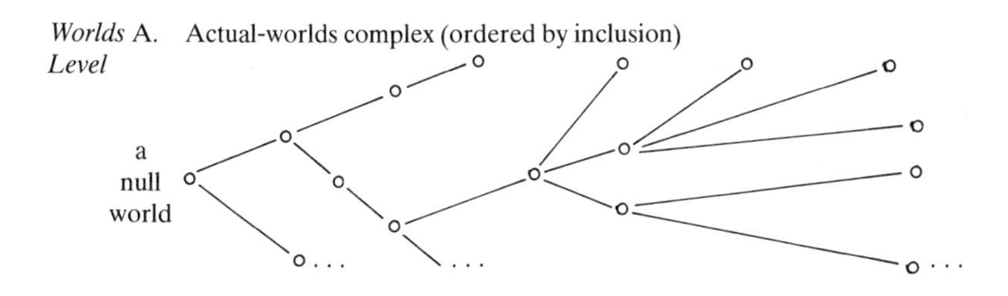

A*. Other worlds (possible, incomplete, impossible, absurd).

It is differences reaching down to the worlds level that are critical. In their more liberal versions, mainline positions can of course tolerate rival positions and a multiplicity of theories, though only a single (unknown) one will be reckoned correct. But mainline positions have now become steadfastly one world. There is nonetheless a long and distinguished procession of many world theories from before Plato through Leibnitz to Popper and beyond. As one example of a better worked out but lesser ranked many world theory, which is important to distance from sistological pluralism, consider Lewis's modal realism. While such realism also includes a multiplicity of worlds, it recognises only *one* actual world, which determines truth, and only a quite restricted set of complete possible worlds (see his PW). These other possible worlds have, moreover, a different status from that assigned to them in sistological pluralism; they are all taken to exist by Lewis, whereas under the chosen position none of these others exist. Thus, although the overall pluralist position embraces a bundle of positions, the chosen position is in direct conflict with other positions in the bundle, for instance over what worlds there are and which exist.

Different realisms themselves supply at least a chain of worlds; naturally each position will reject, or aim to reduce to their own, the worlds acclaimed by other realisms. Contemporary scientific realisms, for instance, tend to be austere, confining what exists to objects taken to be essential for science, by contrast with older realisms which often included God and a richer structure of religious objects and universals. On austerer positions, features displayed in richer worlds, such as values, commonly go the way of relations under idealism, the projection way. They are said to be projected onto the actual world, not part of it. Plainly, by varying the mix of projections with actual world ingredients, a multiplicity of worlds can be distinguished.

What is said of *worlds*, and of their multiplicity, is in certain respects like what has been said of theories and systems, and still more, of conceptual frameworks and paradigms. So an early, and continuing, accusation is likely to be that radical pluralism confuses worlds with conceptual schemes or theories (and actual worlds with correct theories?). But there is no confusion. The usual differences, many and striking, between theories and worlds remain — except that (actual) worlds are not unique. Theories are essentially propositional in character, comprising systems of propositions closed under relations including, always, some form of deducibility. They typically include propositional components such as terms, which are about things in worlds. Theories are presented sententially, maintained, adjusted, derived, and the like; worlds are not (significantly), but are resistant to such manipulation. Though they may be described, worlds are not propositional, but rather factual and thingy, and enlarge upon features of classical model structures. Creatures and people inhabit (some) worlds, but not (significantly) theories. Stars are born and mesons decay, micro- and macro-processes occur in worlds, but again not (significantly) in theories. And so on. Yet the differences are not quite as sharp and robust as realists tend to suppose. Most of what is said of [correct] theories can be roughly transcribed in terms of (corresponding) [actual] worlds, and vice versa. Moreover there are some significant influences and leakages across the world/theory divide.

Because of its multiplicity thesis, radical pluralism differs not merely from definite realism, in its many and varied forms, but also from a standardized opposition, anti-realism and relativism, forms of which also typically assume at most a single actual world (which may however be "internal" or a social construction or an arrangement of "lesser realities"). These types of differences are summarised in the Diagram 2 (standardized parts of which gain fuller discussion in Woolgar, p. 243 ff.).

The realist assumption has generally been that the theory to world relation is a function, a many-one relation, with but one world correlated with many (competing) theories, and *that* world determining (e.g. by correspondence, reflection) the correct theory among them. Such a functional relation is also assumed by relativism, which accordingly has to reject a correspondence theory. Why? This functional relation is not seen to be a matter of great good fortune, or else as something fixed or designed (by a power or god who can choose), or as chosen.

The same surprising uniqueness assumption has been taken for granted over the generations by power-elites and colonisers and missionaries, who have automatically assumed that *their* theories correctly represented the real world, how things really and uniquely were. Their superior technology supposedly helped show as much; at least their magic did yield a military technology of some power. Always their assumptions were mistaken, hindsight is now supposed to show unerringly and with unperplexing regularity. Moreover, other magic, than that of the false[6] combinations of christianity and classical physics relied upon in modern times, could have yielded equally devastating technology. A similar fallacious inference from the technological success of contemporary science (in military wares, space transport, nuclear reactors, and the like) to the validity of the Western scientific view is all too commonly encountered.[7]

The comparison of realism, which is typically transcendental, with religion grows in force when the question of arguments for the uniqueness of actual world is raised. For often the uniqueness theme is simply taken as an article of faith; no arguments or evidential data are adduced. What argument there is gets directed to the existential claim (much of the enormous realist activity is of course polemics and

Diagram 2. Main styles of positions relating correct theories/accounts/documents and actual worlds/realities.

Styles of positions	*"True" theories*	*Inter-relations*	*Actual world[s]*	*Further worlds*
1. *Realism* [Reflective]	One true theory □	correspondence to	Unique existent world □	
1a. Austere realism	no other worlds.			
1b. Modal realism	adds to Realism certain			Possible worlds
2. *Relativism* [Mediative]	Many interpretations (or theories), none ranked ahead □ ⋮ □ □	interpretation of; mediating (social) conditions	Unique existent world □	
1a. Austere relativism	no other worlds			
2b. Modal relativism	adds to Relativism			(Possible) worlds
2c. Cognitive sociology (similarly "orientational pluralism")	like Relativism, but may reject the no differential ranking theme			

3. *Anti-realism*

[Constitutive] Anti-realism reverses priorities with a special theory (or in relativised versions, theories) coming first, and a world — if any (on null versions, there is none) — being constructed therefrom:

One correct theory □	from which is constructed	Unique (internal) world □

4. *Radical pluralism*

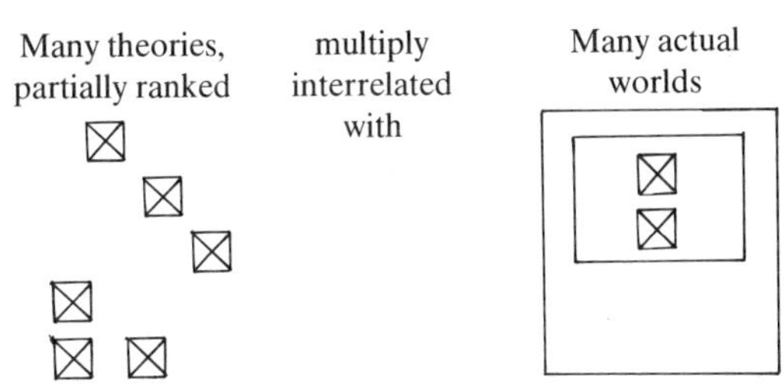

4a. Sistological pluralism *adds* a rich variety of further worlds, and perhaps *subtracts* many ontological assumptions.

apologetics, not sustained argument at all — as will be alleged of the present exercise by irritated realists).

2. THE SUPRISING LACK OF ARGUMENT FOR THE UBIQUITOUS UNIQUENESS ASSUMPTION

Consider first the failure of realists, and others, to adduce arguments for *the* actual world. If we examine realist arguments for [the existence of] the world, such as Moore's celebrated proof of the external world, what do we find? In fact virtually none of the requisite argumentation, but mostly scene setting, which does not even address, however, certain crucial issues, such as what counts as a *world*. The core argument can be contracted into the entailment:

M: I[Moore] have two hands → There exists an external world.

Observe, firstly, that this in no way establishes uniqueness. Observe, secondly, that unless there were something very special about external worlds, uniqueness would *not* follow. It does not follow in general, it does not follow for people or for hands — after all, a rather more convincing argument is the plural ontological one.[8]

M′: I[Moore] have two hands → There exist [two] hands.

Yet Moore simply assumes uniqueness, without adducing any such special features of an external world. Observe, thirdly, that the entailment is invalid, indeed conspicuously invalid once the immense (inferential and topological) distance from isolated objects, like hands, to worlds is appreciated.

Moore's enthymematic argument can be filled out. For example, Moore might have gone on to record, and sum, many further features of his physical environment than those he listed. That would establish, in the ordinary way, only the existence of Moore's environment — of an *umwelt*, to import a convenient term for that fuzzy local object. But an umwelt is not a world, and doesn't guarantee a unique world. (The missionaries and the natives *may* stand in the same restricted umwelt, but they belong as insiders to very different worlds, one for instance with a single God in its existence domain, the other with many or no gods). An umwelt, a subworld neighbourhood, is *that something or other out there* that ostensive methods can more or less mark out. But many and varied are the extensions of a neighbourhood or umwelt to spaces or worlds, as topology demonstrates (this is a main basis of underdetermination: cf. PP).

As it happens Moore, despite his acclaimed 'proof of an external world' (the title of his famous paper), alludes only once to 'an external world', and does *not* there assert uniqueness of it (p. 148). However the equations he makes commit him to uniqueness. For he identifies 'an external world' with 'external things' (or 'the existence of external things', p. 149), and these with 'things outside of us' (see, especially, p. 127); and these things, *the* external things, are supposed unique. In fact, the equations are unsatisfactory; for a world is a whole and involves more than the sum of the things which are its parts; specifically, it involves at least relations, order and structure.

The important point for present purposes is that Moore's hand-waving proof that there exist *some* external objects — whatever the proof is worth[9], it is all he initially claims (p. 148) — goes nowhere towards establishing, and indeed does not really address, the character and disposition of *all* external objects, which help make up and delimit an external world. *Some* external objects, such as Moore locates, are compatible with, and can be fitted into, infinitely *many* external worlds. Moore's argument contributes nothing, then, towards showing uniqueness; there is no relevant argument there.

Impression tends to dominate idea and argument in this area. There is the widespread (but culturally inculcated) feeling that there is just one great mass of stuff out there. How could that neat and singular environment seen out windows or tidily encapsulated indoors as out-of-doors (by people who dwell in solid buildings) be part of several worlds? The idea is outrageous! The impression typically has a verificationist basis. Such is the raw-feeling approach, that the mass of felt data forms a whole, which is single. No doubt it is, single and different for different people and creatures, different generations, different cultures. What is perhaps worse is that the mass of data does not yield a consistent world, or necessarily anything close to what the world is taken to be, even by the source for the data. The mass of data received by any group is characteristically inconsistent, as well as manifestly fuzzy and incomplete; it does not supply an actual world. To be applied the "information" is subject to a considerable amount of filtration, adjustment, verification, classification, etc. But the results of this processing, required to reach an actual world, are hardly unique.

A reason sometimes offered as to why realists and other monists haven't argued for the uniqueness claim is that it is obvious, just obvious once the hard work of reaching the existence claim is accomplished. But it isn't *that* obvious (or an 'unassailable instinct': Bloor p. 143). For after all, it isn't an observational matter, it doesn't follow directly from experience, argument doesn't establish it immediately (or at all, as counter models will show): how is it obvious? What, early on, begins to make it look decidedly less than obvious is the observation that different realists arrive at *different* actual worlds. For example, physicalists arrive at a stripped-down physical world devoid of secondary and tertiary qualities, while commonsense realists point to a richer or commonsense world, of one sort or another, and these days environmental philosophers allude to evaluatively deep and rich worlds which are saturated with value. It is becoming notorious that there are severe problems in reconciling these worlds, because they are incompatible in various important respects. Consistent worlds cannot both contain colour and not, be purely extensional and intensional as well, exhibit value as a genuine feature and not. Nor is it just the incompatibility of these types of realism that suggests that uniqueness is illusory; it is the inconclusiveness of the dialectic between them.

The sheer *extent* to which the worlds concerned differ serves to demolish the idea (floated but never established) that the worlds involved are all somehow equivalent (or its linguistic "equivalent-descriptions" variant).[10] Some of the (less admissible) worlds involved are too poor to admit even approximations of features of the other worlds, without much resisted enlargement of their apparatus and technology: a desert can not sustain a rainforest, a purer physical world cannot deliver deeper environmental truths or even commonsense.

The actual worlds of scientific realism and of local commonsense realism are different. In scientific realism, the world is the way science (in fact dominant science) depicts it; but under commonsense the world is the significantly different and mostly richer way local commonsense represents it (cf. Blackburn p. 45). Brasher scientism, overconfident (given its own historical record) in its own correctness, rejects commonsense as false. But many are the attempts at reconciliation, and, succumbing to reductionistic forces, reduction to one world. A favourite ploy among scientists (e.g. Davies, and Feynman quoted therein, p. 223), borrowed from philosophers, adopts a levels-of-descriptions theory. But the strategy quickly begins to fail under pressure as to which descriptions are *correct*: scientific, commonsense, or other, given that, unmutilated, they reflect different worlds. Only under a generous science, which unlike physicalism doesn't pretend to exclusive correctness, is some reconciliation feasible.

It is however occasionally contended that, despite the apparent tensions, commonsense is after all compatible with physicalism. Like its isolated predecessors, Devitt's supposed reconciliation of commonsense realism with a physicalist standpoint (pp. 69—71) is substantially illusory (no doubt the illusion has had some limited success). His "commonsense realism" is rather like "Soviet democracy"; it is a narrow form of scientific realism with main commonsense content removed but with a certain surface veneer left. Devitt suggests that it is *not* to take a stand against commonsense to allow that 'the commonsense view of stones, trees and cats is error-ridden' (p. 69 rearranged), or to 'view a commonsense physical object as a system of unobservable particles that is wrongly thought to have secondary qualities' (p. 69)! (The very brief argument to compatibility that follows is fallacious, and involves importing irrelevant essentialist assumptions). But it is a stand against commonsense, on commonsense perceptions; and Devitt reluctantly concedes that commonsense realists would sense some loss if they were bullied into admitting that commonsense objects can lack such familiar features as colours and other secondary qualities. So he tries to buy them off with the offer of reductionistic accounts of nonphysicialist qualities. For example, 'according to [his] account, an object is red in virtue of its power to have a certain effect on *humans*' (p. 70)! This shoddy account of redness — or redness-for-humans, as it quickly turns out — is unacceptably speciesist, has to be patched up by the usual open-ended run of further qualifications about normal conditions, healthy receptive observers, etc., requires the correct *effect* (*sensing redly*, in a way appropriate to the species), and *removes* redness from objects. But it is not the conspicuous inadequacy of such accounts (with their typical yet-to-be-supplied details) that matters so much, as the style and result, which are antithetical to commonsense. Surfaces are no longer really red, with the colour in the surface; no, a complex physical process gives the impression — illusion almost — of redness in the surface to a suitably-placed human observer. Uncorrupted commonsense is not so reductionist, and not so cheaply bought off.[11] Granted however commonsense can easily be led astray, as students by lecturers; still it is quite enough for pluralist arguments that a *main* version of commonsense is nonreductionistic and incompatible with physicalism.

3. THE FAILURE OF WHAT CONSIDERATIONS ARE PRESENTED FOR UNIQUENESS OF ACTUAL WORLDS

Furthermore, what arguments are sometimes indicated by realists and others for uniqueness, fail to clinch uniqueness. In the first place the impression is given that certain logical operations — taking unions, maximalising arguments, summation of individuals, or the like — will determine a suitably unique result. But they won't; for instance maximality on its own frequently will *not* imply suitable uniqueness. In most cases the result delivered is not a world, but something else, such as a set; and, even where the objects delivered are not of the wrong sort, the requisite *structure* for a world is lacking. For example, a mereological sum of things has no special structure, but typically admits of many different structures; thus uniqueness is lost as soon as requisite structure is imposed or located.

There is another major problem with such constructional approaches as well, apart from a multiplicity of different constructions available, namely that firm, incontrovertible starting points for these operations or constructions to build upon are usually presupposed: for instance, absolute basic truths or hard facts or, differently, sense-data or things-in-themselves, of some brand. Such supposedly solid starting points are increasingly in doubt. For the starting points are themselves not independent of a certain damaging amount of theory, and are perhaps not duly independent of the world they are supposed to be delineating.

Of course the drive for uniqueness can be satisfied, after a fashion. For it remains possible to define maximal unique items in terms of the pluralistic theory advanced. For example, the complex structure of worlds, which contains (perhaps just) actual worlds as components, is itself a single item. Let us call this object a (world) *complexity, WC*. It is *not* itself a world, since lacking many of the features that make for a world; it lacks the right structure, suitable domains of objects, etc.; it will not even determine truth with sufficient univocality. Nor is *WC* real, but a mathematical object; so (as JB explains) it does not deliver some sort of super-reality (or Absolute). Moreover, it too will vary from pluralism to pluralism, and will be omitted in some because of its paradoxical features.

Almost as pervasive as the assumption that there is a unique actual world out there with all and only the right properties, is the empiricist idea that there is a unique information source which carries appropriate features. Certainly loosely associated with *WC* are various other seemingly unique theoretical objects, such as total available information, defined for example as all the information supplied by the worlds of *WC*. (It is said 'defined *for example*', because other definitions could be offered.) This information *I'WC* – presumably some sort of propositional aggregation; it admits of various representations – will not however be consistent. One world in *WC* will supply the information that *p* (e.g. that tree is really beautiful; that system, virus, planet is alive; Disjunctive Syllogism is invalid); another will supply the information that not-*p* (e.g. because nothing, certainly not those piles of cellulose, is really beautiful, etc.). The informational aggregate of the complexity will, in several respects, resemble the Absolute, now construed as a totality of propositions (not as an organised world of things).

Certainly various filters can be imposed on the information getting into *I'WC*, to remove noise and rubbish at least. Even correctness filters can be envisaged, to filter out *correct* information. But, once again, uniqueness will not be obtained. For there are various rival ways of selecting sub-classes with the "right" characteristics, consistent sub-aggregates for instance. The situation with set theory affords a useful analogy. There are various different ways of cutting down the inconsistent totality of sets of naive set theory, an evidently correct inconsistent theory, to provide consistent set theories. None of the resulting multiplicity of consistent cut-downs is particularly satisfactory, though mathematicians have come to live with certain of them (much as people may cheerfully put up with bad conditions). All are incomplete, and omit apparently correct information about sets that other theories may include, and so on.

Alternatively, *information sources* may be construed in non-propositional terms, as the stuff out there that provides the source of all human and creature perception at least — and emotion? and value? In fact it has to be understood as rather a *potential* information source, or as an information source *in principle* for perceivers, emoters, and so on, were (though perhaps physically impossible) any about; otherwise it will correspond only to a partial world or worlds. Now what is this kind of information source like? If descriptions are to be kept duly neutral, without prejudicing the issue as to whether some of the stuff is of this or that specific sort, then the source turns out to be decidedly noumenal.[12] And otherwise non-uniqueness quickly sets in; there are various information sources, much as there are various actual worlds.

Even if a unique construction with suitably unique result were devised, the question arises, as for some given set of foundational points, why adopt that or these rather than another or others? But if another can be chosen instead, the uniqueness achieved by such a route is cast into doubt, or even destroyed. Yet surely there is nothing sacrosanct about one construction or method or one group of foundations: there is no one correct way (for instance, to Mecca, if that's where we're going), is there?

More influential than logical and informational constructions in supposedly illuminating the path to a unique actual world, was the example of logic itself. There was just one correct logic. Thus logic provided a *method, the* method, for determining all truths in its sphere, and so determining *that* part of the one actual world definitely. And if the sphere of logic was limited, if logic did not determine mathematics and science (as the stricter rationalists thought or hoped), then logic and mathematics together provided an adequate method, at least according to the 17th century rationalists — a single correct way leading to a unique actual world. (A unique world which happily, and necessarily no doubt, coincided with God's *choice* of the actual world as the maximal, e.g. maximally *good*, among possible worlds.) The empiricist opposition did not dispute the *kind* of result, only the adequacy of the class of methods, deductive ones, regarded as exhaustive by the rationalists. But once the appropriate class of empirical methods — whatever they might be — was duly adjoined, then science did, or would, uniquely determine how things were, the unique way of the world. Science (beginning with the pure sciences, logic and mathematics) tells it like it really is — or at least the *natural* sciences do. That was the virtually universal assumption: it also was the main problematic for philosophers, such as the German idealists, who wanted room left, or made, for human freedom, freedom of choice and action, and scope available for morality.

But the basic assumptions were wrong; there are no such unique correct methods. Methods split asunder, and with the splitting, the problematic of 19th century German philosophy fragmented and largely disappeared. There is no one correct logic, no one correct mathematics, or a single set of mathematical methods, no one correct science, or definitive set of adequate scientific methods. It is not merely that the present generations of researchers is very far from having a grasp of such methods, or even of logical methods. There are competing methods and ways, rival *correct* systems (intensional of various sorts versus extensional, for example, inexistential opposed by existential, etc.). Moreover, there are no exhaustive methods; researchers are not going to break through one day to definitive methods.[13]

None of this is incompatible with some sort of convergence, with a pruning of admissible theories as sciences advance, with improved rankings, even with progress[14]. However the convergence is not upon a unique end-point, but towards or upon a perhaps expanding cluster. The convergence theme, commonly taken as part of realism, or coupled with it, is pluralised also. What is modified is the single-valued convergence theme: that by progressive elimination of false views there is convergence (in science at least) to the one true position. There is quite insufficient evidence for this, as well as increasing evidence of scientific failure and deadends. There are other problems also; for example, the "convergence", which is not particularly well defined, is combined with evident divergence. While some positions are assigned to the historical scrap book, or heap, another conspicuous feature which tends to counteract convergence is exponentiation, a stunning growth in the number and variety of positions. For example, one solution tends to generate many new problems, leading to many new positions. There is no compelling reason so far for expecting such multiplication of new and rival theories to end; for there may always be further depth to penetrate. The idea of sharp convergence tends to rely upon the underlying assumption, hardly well-founded and what is now at issue, of a unique end-point to which that convergence is, however erratically, directed.

A unique underlying world is likewise simply *assumed* in the *many views* (or many theories) *but one world* picture much favoured under the dominant paradigm. It is a commonplace of much more philosophical literature from contemporary "science" that, though there are many theories and views, these all give a perspective on a single object, one world or reality. Thus even Bohm, who for the most part sees himself as presenting a non-standard position, insists, 'Rather, all our different ways of thinking are to be considered as different ways of looking at the one reality . . .' (pp. 7—8). (Similarly Davies, Feynman, and a grand procession of philosophers from Nietzsche on.)

How do the many views (dominant science, deviant sciences, commonsense, deep ecology, etc.) *fit together* with one world? Is there perhaps a (suitably singular) procedure, given many theories, for constructing or designing a *single* world which they are all theories or views of? No. There are *various* models and pictures for how the views fit together — so yielding *different* worlds. The most common of these is the *perspectival* picture, a geometrical analogy. Like views of an object from different angles, theoretical views are views of "the" world from different perspectives. But what corresponds to perspective or angle here? The analogy quickly begins to break down[15]. Moreover, the views to be reconciled, by a perspective parameter, are often inconsistent (under one perspective electrons are particles, material objects are coloured, etc.; on another electrons are not particles, but e.g. waves or wavicles or only constructs; material objects are not really coloured but have only primary properties; on yet another material objects disappear entirely; etc.). While a perspective parameter can indeed remove inconsistency, it does too much; it removes the world it is supposed to be ensuring, and how things really are in themselves, as well. Any perspective threatens to be cancelled by another. Unless some perspective gives the "right" view, or a distinguished part of it, there is no way of getting at (or to) how things are in themselves. "Objects" become radically incomplete, tending again to noumenal.

The famous Blindmen and Elephant parable offers a different model of integration from the perspectival, a "multifaceted reality" view. The model depends, however, on the mutual compatibility of the different descriptions given by those sensorily deprived men of the single Elephant. But often, as with quantum theory and real-world theory, the descriptions involved are *not* conveniently compatible, or merely incomplete on one (sense) dimension. There is no further sense available to us, or (possibly) round the corner, like sight, which will give us *the* correct view, of the Elephant-World, denied to the Blindmen — no "reality penetrating" infallible sense, disclosing a further dimension, which will inform us directly how things really and uniquely are, or giving access to a higher-dimensional reality that eluded the venerable ancestors.

There is, furthermore, no single evident trick which fits the different incompatible descriptions — of the "one" world? — together, or which enables recapture of that inaccessible world. If there were, it could be expected that some of the conceptual difficulties of quantum theory would also be resolved along those lines. The descriptions, of different theories and cultures, cannot simply be combined, without adjustment or deletion, else the result is inconsistent, and damagingly so. For apparently consistent objects, such as well-behaved material medium-size manipulable Newtonian objects, then emerge as inconsistent. Nor will taking sets of descriptions do. Once again, sets are not worlds, and cannot provide an account of how the world uniquely is. Deletion, which is regularly required to restore consistency or coherence, is notoriously difficult to apply in a way that ensures uniqueness. To regain uniqueness under subtraction operations, choice is characteristically required.

4. ON ROUTE TO MORE POSITIVE ARGUMENTS TO NONUNIQUENESS

So far the radical enterprise has mainly consisted in demolishing arguments and considerations in favour of uniqueness of an actual world. Realists, who imagine that the onus of proof falls upon any opposition, will be looking for some more positive action. This too can be accomplished. But it is not pretended that any of the arguments and considerations advanced are, or can be, absolutely conclusive (that idea too is a realist illusion).

Initial arguments for a plurality of actual worlds come from phenomenological data: firstly, experience (as an outsider) of different worlds from those of dominant paradigms, and secondly, observation (mostly indirect, where other cultures are involved) of other experience of different worlds. For example, for those who have not contracted the extensional existential disease, mainly a Northern NATO phenomenon, the actual (encountered) world is not restricted to extensional features and its domain of objects is not delimited to those that exist. Indeed many of the objects of dominant positions (such as ideal mathematical and theoretical objects) will be seen to belong to a grander inexistential domain. Those who have escaped infection live in a much richer world. But nothing stops them recognising that infected philosophers can get along, more or less, somewhat as inhabitants of bleak city environments can often get by without breakdown; and they have various crutches and compensatory devices, in the form of reductions, projections, and the like, and at worst they can (and do) discount unfavourable information, for instance, as anomalous. So they (like other one-worlders) cannot be "proved wrong". Nor is it just that such different accounts may be correct; it is what they are *about* also differs irreconcilably. The differences do not end with theories; different worlds are involved. Such differences are sometimes linguistically reflected, in 'our world', 'their world', 'real world' and so on; but such popular discourse even linguistic philosophers prefer to dismiss.

Phenomenological considerations and anecdotal examples get solidly backed up by sociological case studies. There is now a wealth of cases illustrating multiplicity, and scope for multiplicity and suppression of multiplicity, in all natural sciences. Many of these cases illustrate well selection and negotiation processes involved in reaching scientific consensus and unique dominant accounts of how things really are. (A worthwhile survey of cases is given in Shapin; see also Mulkay, chs. 2 and 3). Often the multiplicity is cast, unsatisfactorily as we shall see, only in terms of multiple interpretations, but often enough it is also expressed, rather more adequately, by way of multiple realities (which too often degenerate however into "constructions"). For, given such multiplicity, there is no plausible stopping point on the way down to worlds.

A main argument for moving on to a theory as rich, particularly in worlds depicted, as sistological pluralism looks to its explanatory power (see further §6). *Part* of this consists in its ability to explain the data and dirt concerning science emerging from detailed case studies in cognitive sociology. As cognitive sociologists have expended much space arguing (when not diverted into defending irrelevant and implausible ideas like the so-called "strong program"), mainline philosophies of science perform exceedingly poorly in accounting for such data. The irony of the situation is that the preferred positions that cognitive sociologists move to, at their best forms of theory pluralism, do not fare too well themselves in the face of the data, soon encountering some severe problems (as Woolgar points out, p. 246ff.):

> ... the proclaimed relativism of the ironist conceals a preferred commitment to epistemological realism. Beneath the differences in accounts there *is* an (actual) unchanging reality .... Alleged variations in interpretation between cultures or across historical situations are premissed on the notion that there are interpretations of the same (unchanging) thing. The apparent relativism of social studies of reality belies their practical commitment to realism (p. 252).

Such troubles are immediately removed through multiple actual worlds. For no longer is there a unique actual world beneath the differences in accounts; different cultures can genuinely have established commitments to, or have selected in customary ways, different worlds; and damaging practical commitment to definite realism can lapse.

Not only studies of science, the advance of contemporary science itself, suggest pluralism. For the transition advocated from realism and its standard rivals, seen at bottom as unique actual world theories, to radical pluralism, resembles transitions in the temporal advance of physics, most notably that from Newtonian to special relativistic physics, but also that from Copenhagen to many worlds quantum theory. Consider Joseph's illuminating account of what he terms the "holistic" special relativistic generalisation of the classical situation:

> What special relativity denied was not the acceptability of the Newtonian system of referents for physical magnitude terms in a description of the dynamical and kinematical features of [a physical] world, but rather the uniqueness assumptions made by classical physics ... the privileged status of a particular set of referents for these magnitude terms.
>
> The same point can be put in a different way. The difference between Newtonian and special-relativistic space-time is that the former requires the *unique decomposability* of the 4-dimensional quantities defined on it. There is only one correct way to decompose the Newtonian space-time interval between two events into a spatial interval and a temporal interval; there is only one way to decompose the energy-momentum 4-vector into a momentum 3-vector and an energy scalar. In special relativity, such decomposition into components can be accomplished in an infinite number of ways, one corresponding to each inertial frame of reference [reality framework]. The special-relativistic postulation of the physical equivalence of these reference frames, and hence of the physical equivalence of the various decompositions into components, is ... the holistic feature of the theory. There is no [compulsory] decomposition; which one to employ is a matter of [choice] (pp. 446—7).

Only at the very end is it important to vary Joseph's story, to modify conventionalism in favour of choice, and to absorb "holism" in pluralism.

The swing now occurring in quantum theory interpretation is less decisive, and less favourable, but worth recording. Bohr's "complementary descriptions" approach to quantum theory, an attempt to reconcile consistently inconsistent views of subatomic objects, encounters the same fate as perspectivism; quantum objects become seriously incomplete, and mysterious. The interpretation served to undermine realism about subatomic particles because it left no right description, no account of how electrons and other microparticles really are. They emerge as rather nebulous objects — which fortunately conform for the most part to certain quantum laws. Rival attempts to provide a realist one-world interpretation of the mathematical theory have rather conspicuously failed (undercut by Bell's and other theorems), though much realist effort continues. Not too suprisingly, then, some attempts to reintroduce something like realism into the interpretation of quantum behaviour dissolve the one world assumption. But the very convenient argument to many actual worlds on the basis of an Everett-Wheeler (*E-W-G*) "many-worlds" interpretation of elementary quantum theory has its weaknesses and may have to be virtuously foregone. The weaknesses of such an argument to many actual worlds are these: the interpretation is but a non-standard interpretation of (part of) quantum theory and is in no way required or enforced; it admits of more satisfactory re-interpretation (so it is argued in CS) in a many *non*actual worlds model; and all the "worlds" involved may be regarded as part of a single "super-world", and have been so viewed,[16] though there remain serious unresolved problems about this combination of "worlds" too. Nonetheless the *E-W-G* interpretation undoubtedly gives a fillip to many actual worlds theories, and to the coherence of radical pluralism.

Alternative world experience also leads into another main argument for nonuniqueness, that from variation. For example, possible worlds can plainly be enriched, in various ways, and continue to serve requisite roles, very likely more adequately; such, for instance, are certain intensional augmentations and enlargments by nonexistent explanatory objects. Once again, deletion accompanied by reduction apparatus (e.g. explicit reduction, supervenience reduction, and so on, for other major philosophical games), will provide further worlds. Several of the "ways of worldmaking" Goodman sketches can, in fact, be so applied — though in a new role, not world *making*, but world variation. The methods (which specify in general non-effective operations) for transforming one world to another, can be applied also to world "restoration". One way this can be done is by reassigning the theoretical objects of defective or defunct scientific theories to a domain of non-existents (this gives a rather new look to some of the history of scientific theorizing.)

The availability of variants, of varying calibre, shows that uniqueness is hardly imposed, or somehow enforced, in the way realists appear to suppose; there is no single gripping definite structure. In fact, some of these who claim to march under a realist flag do not really operate as if there were such a definite structure out there which determined every thing (for instance, Australian realists such as Hooker deliberately leave many things open, which however a thorough-going realism would be bound to close.) The arguments from variations and variant extensions can, furthermore, be strengthened in very interesting ways.

Suppose, contrary to radical pluralism, there were a unique actual world. Then there would be a single ideal theory corresponding to it, under the correspondence theory of truth. But this ideal theory admits rival nonstandard modellings. For example, if it is a first-order theory, it will have (by Skolem-Löwenheim theorems) denumerable, extensional modellings. Such modellings determine worlds distinct from the usually assumed actual nondenumerable world. Some of those modellings will supply, as in the first-order case, worlds which are subworlds (or superworlds, perhaps with additional elements) of the assumed actual world. So an actual world is, contrary to assumption, not unique.[17]

There are related semantical arguments for pluralism from extensional (and existential) reduction. Roughly, the theory and worlds of one theory can be modelled in terms of the worlds of another, preserving requisite semantical features; so there is a semantical homomorphism (this connection offers accordingly a correct semantical analogue of the positivistic "equivalent descriptions" theme). Thus, for instance, an intensional theory can be remodelled extensionally, a physicalist theory can be assigned a mentalist world structure, and so on (see SM for details).

Many arguments to nonuniqueness are arguments to nonuniqueness of a *correct comprehensive theory*, not to that of an/the actual world. That is, these arguments, from features of theory variation and undetermination, purport to show that a single correct suitably comprehensive theory is not given or determined:

(A) *there is no such single correct theory.* But it is a big step from theories to worlds, from no single correct theory to no unique world — or so we are told. It matches that, equally large gap, between knowledge and truth,[18] knowledge bearing of course on correct theories, truth on "the" actual world. As luck would have it, however, there is a well-known correct theory which bridges the not-beyond-bridging gap between theory, or knowledge, and world, or truth, namely a generalised correspondence theory of truth. Under correspondence,

(B) *correct theories correspond to actual worlds.* There are several ways this correspondence can go, for instance, it can run from correctness to facts (e.g. correct statements correspond to facts[19]), directly or through truth, or it can connect correctness isomorphically with situations, or the like; and then it can go from facts or situations to some composition of a world, because (e.g.) the facts delivered by a correct theory serve to delineate an actual world.

From (A) and (B) taken together follows the intended conclusion, that

(C) *there is no single actual world.* The connection between (A) and (C) is of course two-way, but has invariably been applied the other way around. It has been argued, especially by realists, that since there is a unique way the world is, there is a single correct view, and (within strict limits, equivalence, etc.) only one, that which describes accurately how things (really) are. But the realist-ceded premiss ('since . . .') is no longer conceded. Of course, the argument, whatever its direction, has often been interfered with by anti-realists and relativists at the correspondence linkage (B); the object-theoretic strengthening of correspondence is intended to stop such damage.

This key argument admits of much variation. For example, 'correct' can be replaced by other (nonequivalent) modifiers, e.g. 'right', 'true'. The circuit through (B) can, furthermore, sometimes be avoided in favour of other assumptions. While certain important arguments for (A) are not readily adapted to support (C) directly — arguments from the nonuniqueness of methods and from theoretical underdetermination, for example — some arguments do admit adaption; notably those from verification and confirmation, and, more solid, those from the role of choice and valuation in extensions beyond whatever is (fairly immediately) given. The crucial role of choice and valuation in determining worlds that are determined means that uniqueness cannot be imposed, e.g. by *fiat*. For choices can be differently made; different values can reasonably be held or adopted. Value enters in several ways: in the methodological criteria involved and their rankings (e.g. how much parsimony counts, whether extensionality is required, etc.), in the place given to explanation (which may be interest relative, and value dependent in this and other ways), and so on.

Nonuniqueness is an immediate outcome of the way the basics of a preferred actual world are *selected*. For this selection settles how things really are there, what is really there, as opposed to what is projection, construction, fabrication, myth. Such divisions are not uniquely defined; many different selections can, and have, been made. It may be objected that an actual world is a given, and that such division is a conceptual carve-up imposed on top of it. A world is not however a given, at most an umwelt is; worlds are theoretical objects, and thus informed by theory (see JB and SF). Nonuniqueness is an outcome not merely of initial selection, but also of *interpretation* and *augmentation*, which can happen in many different ways. Augmentation proceeds by the addition — location, discovery, invention, construction, shaping, forging — of theoretical, explanatory objects, both of theoretical science and of mathematics. Many of these objects are — or prove to be — nonexistent.[20] By virtue of the formal elaboration and mathematisation of theories, "there will always be" a heap of these items.

A main route to (C) does proceed indirectly from (A) through a different elaboration of (B). Most simply, (B) can be rendered *analytic* by a suitable, and natural, account of "actual world". The question of what counts as an actual world is of course decidedly critical, not only for radical pluralism, but also for various theories that want to insist upon a single actual world or single reality. In the first place, 'actual' modifies 'world'; actual worlds are those items among worlds which are actual. Thus (much as strict empiricists may protest) the labour may be divided in two tasks, that of distinguishing worlds, and then that of providing criteria for determining actual ones among them. It would, by the way, be somewhat surprising if a test for a property — the Kantian feature of actuality in this case — inevitably resulted in there being exactly one of the sort; there would be some suspicion that things had been rigged.

Though notions like *world, cosmos, universe* have bulked very large in the history of philosophy, there is, unfortunately, no ready-made account of *world* that one can simply appeal to or commandeer (despite the enormous contemporary screed on worlds from Heidegger through Lewis[21]). Perhaps the best accounts to build on come from object-theory literature (we shall take over with but little adaption that in JB, p. 202ff.). A world is an object, a (dynamic) theoretical object, of a certain rather complete sort; it has, like a logical model, certain domains (of objects, etc.), it stands in certain relations, etc. (for relevant details see JB). It also has a certain structure, an order, coherence, suitable comprehensiveness, system of classifications, and so forth.

Still more crucial than the matter of what a world is (about which a certain nonchalant vagueness can be got away with in many of the arguments advanced), is the question of when a world is actual. There is a straightforward answer, which guarantees (B), virtually by definition. A world is *actual* when it corresponds to a correct theory on how things are; that is, a world is actual when it represents *a* way of things correctly. Thus, actuality does not effect unique selection among worlds, because (and *as*) correctness does not among theories. Two important features of actuality are thus invoked: that being actual does not guarantee uniqueness, but sits with multiplicity (as in 'actual days worked'); and that, as applied to items such as worlds, it ensures correctness. Both can be argued for on the strength of the meaning and use of 'actual' as supplied (not altogether adequately) from dictionaries. Firstly, 'actual' certainly permits plural couplings, and nothing in its usage excludes plural combination with 'worlds'. Thus uniqueness of an actual world is not analytic on the sense of 'actual'; such cheap routes to uniqueness break down immediately. Even on the unlikely view that 'actual' is a demonstrative (implausible because the essential pointing function is frequently excluded, especially as regards worlds) or Lewis's dubious view that it is a purely relative indexical (dubious because it doesn't operate in the fashion of paradigmatic indexicals), plurality is perfectly possible; consider 'these', 'those', 'we', 'their'. Indeed Lewis's analysis wrongly makes 'actual world' work like 'this world'; both the analysis and the relativity claim neglect the further sense of the term. Secondly, 'actual' is associated with such adjectives as 'real', 'existing', 'current', 'present' as opposed to 'virtual', 'theoretical', 'imaginary', 'potential', 'possible'. It thus provides a way of things as they really are now, a way of a world, so guaranteeing correctness of corresponding theory.

5. ENCOUNTERING OBJECTIONS TO CHOICE OF SISTOLOGICAL PLURALISM

There are various logically-based arguments designed to show the incoherence of pluralism, and thus to force uniqueness indirectly, to collapse pluralism to monism. Most of the arguments to inconsistency or incoherence depend upon confusing pluralism with relativism, and whatever their plausibility against relativism, they do not extend to apply against pluralism, which *can* discriminate among and rank positions, theories and worlds. Pluralism does not face such relativistic difficulties as that no theory is better than any other, though relativism — itself a theory — is preferable or even correct. For the egalitarian-looking premiss is not granted.

It is important, then, for the coherence and intelligibility of pluralism that it be been clearly distinguished, and distanced, from relativism. Yet what is more common than to conflate pluralism with relativism (thus e.g. Berlin, Rescher, and many others), or to call what is pluralism relativism (thus e.g. Goodman)? What they do have in common is both a multiplicity theme, many theories or many worlds, depending on the level, and a qualified rejection of absolutes, absolute unsystematised truth for instance. But that, and opposition to monistic rivals, may be about as far as the overlap goes. Relativism implies relativity to something, a relating; pluralism doesn't have that construction (though it does of course involve some system relativity, commonly contextually supplied). Relativism presupposes no ranking of positions or else equality of ranking (and sometimes even that "anything goes"); pluralism certainly does not. The point is especially important in avoiding familiar self-refutation charges, lodged against relativism, which do *not* automatically extend to pluralism (because e.g. "the law may be laid down" from a much preferred position). The reduction argument against relativism starts from this sort of charge; relativism is itself advocated as a preferred position when no position is preferred. These forms of self-refutation do not touch radical pluralism. Nor do others. For a logical modelling (elaborating that of PP) enables demonstration that sistological pluralism is substantially immune to damage from self-refutation attacks.

A further supposed difficulty for pluralism, linked to self-refutation, is this: How can we explain pluralism, its different worlds, *unless* we do so from a common world? The answer is: from one world or other, or overlapping ones. Some will do; a unique one is not required.[22] In the end, radical pluralism, by contrast with relativism, need not deny that a single world can be selected and an actual world position more or less consistently worked out from there — many such positions can, with different select actual worlds. The process involved, and supposed results achieved, are what are at issue: the claim is that the choice of world is not forced, but can be, and sometimes is, made differently. Different choices will lead to different classes of problems and solutions, etc.

Normally a culture does distinguish and adhere to a single world, *the* actual world in the context, i.e. in the cultural setting (to make contact with the contextual definite description analysis of JB). Alternatively, if the society is complex and more pluralistic, there may be a small range of worlds, one of which each different sub-culture distinguishes. For example, in Northern industrial cultures, there is the (scientific) world of scientific elites which is like dominant science tells it; there is the (commonsense) world of everyday life, which is like common sense says it is in outline; there are various religiously suffused worlds, under which . . . ; and so on. What the dominant groups of sub-cultures accept does not, however, have to be agreed to by an individual. An individual can make a different, not perhaps idiosyncratic, selection of an actual world, or can have come, without any conscious selection, to have accepted such a world.

Mostly, of course, such cultures and individuals do not select their accepted actual world, they inherit it, are educated or brainwashed by reference to it, and so forth. Still less is the actual world they discern, and live in, a conventional choice; concerning worlds there is little organised choice, and but few conventions. In certain respects, then, 'choice' is an unsatisfactory term in this context; for choice may not operate, as with inheritance, or may be submerged, as with conversion, where a switch of worlds may be affected. A better term may be 'latent choice', because choices perhaps unrecognised have been implicity affected, and because the potential for choice is available to suitably-placed free nonsubject subjects.

Pluralism is then not the same as, and not even very like, relativism. For pluralism rejects such relativistic assumption as that anything goes, that nothing is inadmissible; it rejects relativistic egalitarianism, and offers instead an account of world evaluation, ranking, and latent selection, which permits nonrelativistic single world selection and adherence. (From this perspective, definite realism has forgotten its mythical constrained-choice origin.) The contrast often made between (egalitarian) relativism and (unique) objectivity is accordingly a false one, because it is far from exhausting viable options. But nor is pluralism itself uniquely determined. There are different forms of pluralism, different frameworks within which a plurality of positions may operate.

Why select sistological pluralism, that is, a pluralism based upon object-theory? A first reason is that some selection is practically inevitable. Life is short and time limited; by comparison information and knowledge are unlimited, and nowadays one can only obtain and retain so much. It doesn't follow that because one cannot obtain it *all* (Faustus fashion) one couldn't or shouldn't aim for a general framework or for *any*, a fallacious inference suggested in some Taoist dismissals of knowledge as an important goal. It does mean some rationalisation (and rationing) of the number and variety of theories pursued. Partly to reduce complexity, then, but also partly because of severe limitations of space and energy, only one main form of pluralism has been outlined, sistological pluralism. It is but one style of pluralism, an apparently novel, radical form. It does not, like wishy-washy orientational or entrepreneurial pluralism operate only at the theory level (or meta-level), somehow abandoning the hard-to-separate world level (or object-level) level to realism and its more puritan rivals. There is excellent reason not to do so. For one, sistological pluralism does not need to rely on an implausible sharp cut-off between levels. For another, it does not simply give a pluralistic veneer to orthodoxy and dogma, such as accompanying object-level "realism" may underwrite.

Practical limitations foster a legitimate tendency to adhere to some favoured position among the many that can be made out, and to treat that position linguistically in a typically absolutist fashion, though its nonuniqueness is not forgotten. Natural language, to which sistological pluralism can stick fairly closely, makes this easy, because of the way it deploys context to help fix signification. Indexicals are simply the most striking examples of expressions where significations are determined in context. Under pluralism a batch of propositional evaluative expressions, such as 'true', 'false', 'worse', behave likewise in a quasi-indexical fashion. It can be plausibly argued that they behave in this way anyway, depending for their assessment on the speakers' select framework.

Rather than alter language significantly then, pluralism *extends* it, where required or advantageous. An example concerns the matter of saying that there is something right, some truth in, all admissible positions: a convenient location is that of pluralistic truth (or '*p*-true' for brevity), an ungraded notion. A proposition is *p*-true if it holds in *some* admissible actual world, whereas it is true (now I'm speaking) if and only if it holds in a suitably rich intensional actual world (vaguely discerned by my mob). Given such pluralisation, further standard objections are easily met or evaded (see PP).

A standard (absolute, obsolescent) conception of truth, or of factuality, is often invoked with a view to enforcing uniqueness. What Descartes claimed is oft repeated, that among the 'many conflicting opinions there may be regarding the self-same matter, all supported by learned people, . . . there can never be more than one which is true' (*Discourse*, pt. 1). But such claims already presuppose that there is a *single* framework, or world, with respect to which truth, "the" truth, is evaluated, and so beg the question against thorough-going pluralism. "The" truth and "the" actual world each pluralise, in correspondence. The standard assumption is not just that where *A* (e.g. "Quarks exist") is inconsistent with *B* (e.g. "Quarks do not exist") then not both *A* and *B*, but that at most one of *A* and *B* (and in the example exactly one of *A* and *B*) is true, true *simpliciter*. True simpliciter takes its place in a pluralist setting, along with *other* truth predicates; it amounts to truth in a certain select world (in Descartes' case, a sparse brutal world, selection of which is by no means obligatory). All apparently absolute definitions of truth, of factuality and the like, involve latent selection, choice of an actual world, a special model — or of a preferred metalanguage enabling infiltration of such object-level choices. In the case of a coherence theory of truth, for instance, choice may be explicit (see CT). As regards a semantical definition of truth, which underlies and can be integrated with other viable accounts of truth, choice is implicit; choice is the only way of breaking out of serious circularity which otherwise arises (for essential details, see JB, p. 330ff.). Such points provide the components for a further argument, from nonabsolute truth, against assumed uniqueness.

A major reason for a sistological choice is of course to avoid excessive ontological claims and commitments. It would be rash to try to assign existence, or some sort of existential status, to the whole colossal pluralistic array, or even to very much of it, and it is quite unnecessary to do so. The reasons are again those of object-theory (in noneist form: see JB and WS). How much of the framework does exist? For *p*-existence, pluralistic existence, the answer can be obtained by looking up the existence predicate in the admissible position concerned. But for existence proper? The answer will presumably be (though it doesn't *have* to be) that almost all, or even all, actual worlds do not exist. What is actual sometimes does not exist! A coherent position does not, and cannot, keep in step with normal contextual implications everywhere. In application of 'actual' to theoretical objects, normal conditions no longer obtain; easy import of familiar ontic assumptions concerning theoretical items does not deserve encouragement. (Alternatively the awkwardness could be mitigated by a terminological shift; e.g. as in some expositions of modal logic, replace 'actual' by 'distinguished'.)

Further reasons for a sistological choice are to resist reduction strategies, aimed at reducing richness, and trivialising strategies, aimed at collapsing pluralism into monism or into incoherence. Many of these moves are engineered through arguments concerning identity, which object-theory is well-placed to fault.

The crucial issue of uniqueness is inseparable from those of identity; and identity *appears* to raise a snarl of problems for any deeper pluralism. There is, for instance, the difficulty, puzzling to some and much exploited by others, as to how worlds can differ, yet may contain precisely the same individuals. The short answer is that one and the same individual may have different attributes (including existence) in different worlds, somewhat as the same individual may have different features at different times (for the rest of a larger answer, see JB, p. 368ff., p. 593). The difficulty is not a special difficulty for a plurality of actual worlds; it applies to plural worlds theory generally, and, despite a lot of fuss, is straightforwardly met in quantified modal logic. Indeed the difficulty, such as it is, is noticeably *less severe* for actual worlds than for merely possible worlds, because the range of *variation* permitted individuals is considerably smaller, though by no means zero. Unlike possible worlds, actual worlds concern not the way things might have been, but a way they are, however they are, whichever they are. For instance, Rangi who in fact has a smooth brown skin, can appear in some possible worlds with a pock-marked white skin, but not in other actual worlds (to be sure he can show up in the preferred actual world of some scientific "realists", not white, but with no colour, indeed without a skin at all). In the main, actual worlds will overlap considerably, and share everyday individuals (overlap, indeed coincidence, of individual domains of worlds is a commonplace of noneccentric modal logic semantics). Everyday individuals will tend to differ not in their surface, observable features, but in their unobservable, theoretical attributes, as in their interrelations with unobservable or theoretical objects such quarks or minds, spirits or Gods.

6. INDICATING MORE PRECISE BEARINGS: WHERE SISTOLOGICAL PLURALISM STANDS REGARDING OTHER FAMILIAR STAKED-OUT POSITIONS

As regards *instrumentalism* at least, the matter is comparatively easy. The pluralist theory selected agrees with instrumentalism that a fuzzy observational/theoretical division and related theoretical/concrete division both obtain. (It concedes however the superiority of a three-or-more way fuzzy classification of objects, distinguishing instrumentally-accessible manipulable objects from purely theoretical ones. Objects may cease to be merely theoretically postulated, and become indirectly accessible; thus e.g. electrons presently, but not quarks.) It agrees with instrumentalism that realist ontological claims, concerning theoretical objects and universals, are excessive and frequently lacking justification. But it disagrees with instrumentalism that theoretical statements are not propositional, and it takes issue with its reductionistic attempts. Instrumentalism is an attempt to escape that usual ontological story, which fails to comprehend the sistological alternative. In large part it results from *accepting* ontological assumptions, failing to notice that people can, and do, talk fairly straightforwardly about theoretical objects, in a propositional fashion, *without* being committed to their existence.[23]

With regard to *foundationalism*, the situation is more complex. In the first place, the building/construction picture that has so often been used has severe limitations. For one thing, there is continual work repairing and adjusting the foundations. These get adjusted *as* the further structure goes up, not simply taken for granted as in standard building. A better image can be developed from the constructions of the Arab river people, who have regularly to adjust or repair their whole floating houses (or, differently, of plants on eroding or flooding ground). For another, different cultures can in principle choose differing styles of shifting foundations; foundations, perceptually-cemented together, do not offer a route to uniqueness. In any case, empirical underdetermination issues, crucial for pluralism as for relativism, are independent of foundationalism. For even where there are secure foundations, the superstructure will not be thereby determined, but only constrained. The theme of the theory-ladenness of all observational claims is not however so independent, but taken to undermine any foundations. Pluralism, while recognising the theory-pollution of older foundations, does not require such a strong ladenness theme; and, except under inadmissibly low redefinitions, of 'theory', the argument to such a theme fails (JB, p. 814ff.)

As regards *realism*, matters are more complex still, partly again because realism comes in so many forms, and has tended to mean all favourable things to too many philosophers. While the *thing*alism (*res*ism) of realism is no doubt alright (and the tendency to static or substantial formulations is easily avoided), the existentialism (*re*ism) of realism is not. A major defect of realism is that ontological assumptions are built into it (as into Fine's so-called "natural ontological attitude"); it proceeds illegitimately, without adequate basis or often even notice, from things to existents. So all sorts of nonexistent theoretical and mathematical things are erroneously ascribed existence.

A corollary is that sistological pluralism questions, on several scores, and assigns a dispreferred rank to, even what has been called (e.g. by Hooker) "moderate realism", 'that there are (= exist) the things that science says there are *and* the world is substantially as science says it is'. By no means all the items introduced, even in dominant science, exist. Nor is there a single science describing a unique world. All these immoderate assumptions of moderate realism are discarded. "Extreme realism", an outrageous scientism, adds to moderate realism a closure clause: and that's all. There is nothing else than what is required by (unified) science, and so ultimately by physics. As the claim is so extremely short on evidence, it adds insult to moderate indecency.

"Scientific" realism is shot through with such extravagant claims: it often reminds outside infidels of nothing so much as a religion; its ranks have been swollen by converts from Christianity, and its theses roll out like a catechism. Like it, many of the characteristic theses of scientific realism (as quoted from Leplin pp. 1–2, and as listed in smaller bundles, from schisms, in many other sources) go down. They are queried, qualified, or scrapped, as follows. A main argument has been directed against the assumption that 'the history of at least the mature sciences shows progressive approximation to a true account of the physical world'. The supposed goal is not well or uniquely defined; and as studies from cognitive sociology are now revealing, that is, by and large, *not* how science proceeds. For connected reasons, it is but an illusion that 'science aims at a literally true account of physical world, and (that) its success is to be reckoned by its progress towards achieving this aim'. Moreover, there is quite insufficient evidence that 'the best correct scientific theories are approximately true'. It is far from certain that 'the central terms of the best current theories are genuinely referential'; many of the theoretical terms do not meet basic requirements for referring to existent objects. It is not always the case that 'scientific theories make genuine, existential claims', and decently reformulated many of the purported existential claims would give way. It is false that 'the (approximate) truth of a scientific theory is the only possible explanation of its predictive success', and dubious that it is 'sufficient explanation of its predicative success'. Nor is 'the predictive success of theory . . . evidence for the referential success of its central terms'; for what does not exist may explain and facilitate the predication of features of what does (see SF).

On the other hand, sistological pluralism is almost, if you really insist, an indefinite realism, ontically neutralised. It accepts much of what realism asserts (at least in surface form), notably that there is an external world substantially independent of minds, perceivers, receivers, and such like. It applauds realist hammering of the human chauvinism typical of *anti-realism.* For very many things are independent of humans, and do not depend on humans at all for their existence (or not) and character. The point holds good not merely for external material objects and other living creatures; it also holds, though differently, for mathematical and other abstract objects. Those mathematical objects that happen to be regarded as interesting and get studied may be selected by some humans, for instance on the basis of their interest or theoretical utility, but they are not, and not thereby, human creations or constructions (see JB, chapter 10). Many scientific realists would prefer to dispose of or forget about these embarrassing objects, some of which play crucial roles in the very best scientific theories; their focus is on material objects, whose very materiality anti-realists would try to dissolve! (Reductions are fine — provided they are of the approved sort.)

As with realism, material objects — of a rich variety of sorts and character — have (mind-) independent existence. They exist, furthermore, in a physical world (i.e. they belong to the existence domain of such a world). And they exist externally to human (and other) perceivers. These themes are often taken to be the hallmark of realism, indeed sometimes to be definitive of it (cf. OED).[24] But they fail to capture current philosophical assumptions, and they are far from definitive unless some, admittedly pervasive but already exposed, false dichotomies are interposed. Otherwise they only serve to mark off realism from a range of subjectivisms and idealisms, not from relativisms and pluralisms (or from a monistic noneism).

To make the requisite contrasts, it is often advantageous to *relativize* realism itself, to replace the absolute term by 'realism about'. "Realist about" sometimes implies "nonreductionist about" (and noneism, virtually by definition, is not reductionist about items), but it also often equates, very differently, with "reist about". In these terms, sistological pluralism is realist about material entities and generally reist about them; but it is not reist about theoretical objects, nor reductionistic about them. While realism sometimes signals transcendentalism, scientific realism commonly involves some heavy reductions (dictated especially by the fashions in logic and physics). Sistological pluralism is also realist about relations as the point is often explained, namely that truths concerning relations between objects need not depend upon truths about particulars which they relate; but it is not realist, i.e. reist, about relations in ascribing them (independent) existence. Relations hold good or not, but that does not entail existence; truth and existence are very different objects. Truth and beauty, goodness and reality, validity and existence, are all different nonabsolutes.

Finally, sistological pluralism offers many of the significant explanatory advantages of *idealism* without the chauvinism and without the serious reductive drawbacks. It can explain why we find "the" world conforms, most conveniently, to various regularities, including those of causal and inductive smoothness, why it exhibits amenable simplicity, and so on; namely because it has been selected to ensure these unexpected but prized features (see further CT and Blackburn). Actual worlds voiding these features have been avoided. It can explain such features, as selection features, *without* however abandoning the claim that there is an actual world extending way out there independent of humans or observers or minds, without reduction of actual worlds (or valuable parts of them) to human constructions or projections or mind-stuff. These are big explanatory advantages, and form a major part of the case for sistological pluralism. They help account for the pursuit of what, to many impatient realists, seems an unlikely, a counterintuitive (not to say, counterlogical), and unnecessarily prolix position. In fact, sistological pluralism is none of these things. Nor are explanatory advantages the only reason for its pursuit.

Justification for looking at, studying, presenting or showing off a pretty logic, or an unusual radical theory, does not have to be of pragmatic cast, that it does something, succeeds somewhere, works well. It may just be a fine enough thing in its own right — somewhat as a tract of rainforest may be valuable in itself, and, though rather rich and complex, not particularly well-organised, or useful for humans, or good for this or that, or unique. Sistological pluralism is a fine and rich theory, fun to work and play with, and to contemplate, irrespective of whether it is good for much — which, as it fortunately happens, it is.[25]

RMB 683 Bungendore
Australia 2621

NOTES

[1] There is some misconception on the part of proclaimed realists about what realism involves, about the need not only for the existential claim but the uniqueness claim as well, if orthodox realist theses are to be derived.

Stove provides one local (Antipodean) example, while Devitt may be another, and Hooker yet another. Laudan appears to be a North American example. But not too much hangs on whether such an overexploited and indeterminate tag as "realism" applies. Call the position involving the uniqueness as well as the existential claim *definite realism*, if need requires.

It *is* standard, however, to regard realism as implying uniqueness. Thus, e.g., Blackburn asserts that Realism is committed to the thesis that there is a *single* way the (natural) world, nature, is (p. 45). Similarly, Fine, in identifying the 'core of realism', claims 'first, realism holds that there exists a definite world' (pp. 51—2). And according to Putnam, metaphysical realism tell us that 'there is exactly one true and complete description of "the way the world is"' (RH, p. 49).

[2] A background modelling is provided by weak modal logics. Such a framework with many actual worlds, called distinguished worlds, may be found in Segerberg for instance. A choice function, such as Hilbert's, can then be applied to effect selection, if desired. Paraconsistent elaborations of such modal modellings are indicated in PP. These modellings both establish a certain coherence and serve to defuse simple self-refutation charges against radical pluralism.

Notice of course, what is hardly news, that many of the key terms — *WORLD, REALITY, EXTERNAL,* etc. — are not particularly well defined. By contrast with the usual literature, *some* of the fuzziness will be removed as we proceed however.

[3] The false contrast is in large measure due to an oversimple background picture, which, neglecting world*s*, yields only the triad of positions — Realism (or Transcendentalism), Anti-realism (Idealism or Phenomenonalism), and Scepticism. The picture found in much epistemology (e.g. in Wisdom's entertaining work), often takes a building outline of the following sort:

Superstructure	Correct theory or position; knowledge claims
Foundations	Evidential base; verifiable claims

The problematic is that the theory, knowledge claims or whatever, appear to exceed their evidential base (which reflects only *part* of an actual world). The realist and sceptic agree that they do exceed the base, the anti-realist denies this. The realist and anti-realist agree that the claims made are justified, the sceptic denies or questions this. And so on.

The outline, in addition to its neglect of worlds, is excessively foundational: it fails to depict the impurity and theory-pollution of the evidential bases, and it fails to take account of the roles of interpretation and selection in transitions from "bases" to superstructures and vice versa.

Much as there are two ways of presenting reductionisms ("there are no *X*s: only, what they reduce to, *Y*s" *vs*. "there are *X*s, but *X*'s are nothing over and above *Y*s"), so there are two styles of anti-realisms, no world *vs* one mind-dependent world. The phrase "null world" compromises between "the" world lost and subjectively regained.

[4] Such human chauvinism is depressingly conspicuous in Rorty and Putnam, to cite just two American examples. It tends to be a feature of the idealist-verificationist-pragmatist bundle of positions that they would dispense with natural worlds altogether (if they could) — a highly negative feature.

[5] This is not a new development, though one delayed through temptation by the illusory ideal of a universal logic — ultralogic — and perhaps universal (and uniquely correct) theories built thereon. A theory of a pluralistic sort was emerging, and is foreshadowed, in JB (a main guide to the notions of this exercise), with its various object-theories and other rival theories, and with multiple actual worlds, especial emphasis being put on worlds *G* and *T*. This pluralistic trend was pushed further in subsequent work, especially PP. A major disadvantage (which there are contextual and other ways around) is the complexity it brings with it.

[6] Where *false*, simpliciter, is written it is contextually specified as false from the adopted position in the pluralistic panorama. The other positions, if admissible, will of course also be p-true; for details see PP. Similarly such terms as 'correct' and 'wrong' are *contextually* relativised, to sistological pluralism.

[7] One conspicuous example is Ziman, p. 9ff. The deductive inference involved in arriving at the validity of an ideology from the (partial) success of its applications or technology is but a giant conversion, of the abductive form $A \rightarrow B, B \therefore A$. Presumably however abductive reasoning, if it ever it attains decent formulation, will, so it is said, serve to show that there is sufficient right about contemporary science to adopt main parts of it as working theory and basis for variation.

It is *not* being insinuated that realists need be dogmatists; only that realism supplies a significant basis for dogmatism. It should be added that whatever the good intentions of some realists, scientific realism and materialism encourage dogmatism, within science and outside it. Of course the objectionable dogmatism of power groups (almost invariably realists) concerns what they take to be the *nature* of *the* actual world, not just its exclusiveness.

[8] For contrast M'', Pegasus has two wings → There exist [two] wings; and compare D', I[Descartes] think → Descartes exist [There exist people]. On the *already* assumed existential loading of the antecedents of such entailments, see JB, p. 27ff.

It is amazing in retrospect that such shoddy argumentation as Moore's helped win the day against the British idealist opposition, then entrenched. A better explanation is perhaps that, as with a political party that has governed for long, idealist forces were exhausted, and the mood in the country was for a change.

[9] A common reaction has been that Moore's "proof" begs the question, takes for granted what needs to be shown against the sceptics and in removing 'a scandal to philosophy'. A rather different, more drastic, reaction is Wittgenstein's, reported accurately enough by Edwards: Moore ties 'the reality of the external world' (it is interesting to observe what Moore was supposed to be about, shoring up the reality of *the* external world) 'to two questions . . . (1) Are there physical . . . objects? (2) Do we know that there are physical objects?' (p. 171). A central part of Wittgenstein's response (the subterranean rumblings of an old defective positivism) is that these questions are out of order, not genuine questions; they are (philosophical) nonsense, reflecting disorder. As the paper indicates, this type of (philosophical) position obtains only a pretty low ranking. Of course, Wittgenstein does much more in *On Certainty* than wheel out an old positivism; e.g. rival views are put up only to be pulled away, to break the grip of defective pictures associated with the Moore's questions, and their answers.

[10] The *claim* of equivalence is that all the (correct) theories — worlds — that remain standing are equivalent, somehow. How? Arguments for the claim are only sketched in particular cases. It has not been established generally, and cannot be. For instance, a more highly intensional theory cannot be reduced to a less intensional one or at bottom to an extensional theory, without the illegitimate introduction of additional primitives. Of course given further primitives, e.g. semantically more worlds, equivalences can be semantically instated: cf. SM.

[11] Devitt's assumption (in later chapters) is that scientific realism, so to say, includes commonsense realism, simply exceeding it in the extensive existence claims it makes (which are however said to be justified through inference to best *available* theory!). But that not only neglects the nonreductionistic character of commonsense, but glosses over its highly localised features: for instance, commonsense is *not* committed to universality, or even extra-terrestrial validity, of its claims (such as that unsupported heavy objects fall down).

[12] The discussion suggests a way of modelling a noumenal "world". A noumenal world is one where the "objects" are highly indeterminate in what properties they have; it is an extremely vague world — an ultravague world. Such a vague world may be represented in a similar fashion to an obvious modelling of vague features. On the obvious modelling, a vague property f (e.g. purplish-brown) is specified in terms of the bundle of exact determinates of f (e.g. all the precise shades under some colour coding between purple and brown). In a similar way a noumenal world is represented through the set of worlds which render it determinate in every (extensional) respect, that is

through a set of non-noumenal actual worlds (or in idealist terms, phenomenal actual worlds).

In a sequel, a theory of the stuff (out there), not an actual world, but unique, will be developed.

[13] These large claims are argued, in more of the requisite detail, elsewhere: on the plurality of science, see e.g. SS and PP; that there is *no* single method, see NL. There is now a rich literature on the nonuniqueness of logics and of logical methods, which included, early on, such pace-setting articles as Waismann's, and now includes social studies like Bloor's.

[14] It is sometimes claimed that pluralism is incompatible with the immense progress that has obviously occurred. It is not. But environmental pluralists, among others, would want to query or reject the enthusiastic picture of progress imported.

[15] On the failure of the perspective perspective, see further PP. In his investigation of orientational pluralism (1978, p. 233) and his inflation thereof (1985, p. 173ff.), Rescher offers us four views of the way theories are related to reality and truth is determined, lumping along with standard perspective and multifaceted ways of trying to accommodate a plurality of theories to a single reality, "no reality" and "unique reality" views. All of these views are attempts to gear theories to a single (perhaps many-sided) reality. Thus major pluralistic alternatives are omitted; and *none* of the views do what is required.

Rescher's own limited pluralism is far from radical. Indeed it does not even extend to "object" theories such as those of science and philosophy, but is supposedly narrowly confined to metaphilosophy. But it is far from clear that it *can* be contained there. The problem is not just the illusiveness of the ever-moving, now it's there now it's not, boundary between philosophy and metaphilosophy. The serious problem is that assumptions easily get over the boundary. For instance, a single godless reality assumption gets to eliminate theistic metaphilosophical positions as incorrect. Rescher has yet to establish the coherence of his pluralism.

[16] Thus, e.g., Barrow and Tipler, p. 476.

[17] This argument, which still contains some fillable gaps, varies an argument Putnam has tried to use, in several places (notably 'Models'), against realism. Some of the restrictions Putnam imposes, or ought to impose given his logical resources, on ideal theories for his argument to work, can be lifted, e.g. extensionality and consistency (nontriviality suffices). It may indeed be that there is no such ideal classical theory as Putnam envisages, but are complete inconsistent nonclassical theories, and various consistent incomplete approximations thereto. On such world-models, see JB (index).

[18] Knowledge used to include truth. Nowadays there is a two-way gap, between received knowledge and truth (and various types of knowledge: reliable, certified, etc.); see further SS.

[19] A suitable correspondence theory, matching facts to truths, is established in SC. The linkages of worlds with theories, and actual worlds with correct theories, in sistological pluralism relies heavily upon correspondences, which also justify some otherwise illicit terminological transference (and give more ammunition for the realist charge of theory/world conflation).

[20] An argument from the history of science, the problem of ancestors, enters here (the problem is neatly introduced in Nola). The problem of *these* false or dubious theories suggests in turn the following meta-induction: items in past theories are now seen not to

exist. It is likely that from a future perspective, the same will apply to some items now postulated to exist.

A partial sistological resolution of this type of puzzle is straightforward: these items did not exist, and should only have been added as purely theoretical objects — perfectly feasible practice — in the first place. More, much more, is required for existence than mere postulation, or, for that matter, holding a place in a presently best explanatory theory. (Some subject terms of science refer, some do *not*. Postulation does not ensure reference, that what is *spoken* about does exist. Nor does explanatory power.) If the requisite further conditions are met, then the problem of the ancestors largely dissolves. For something existent will remain, even with the supersession of the associated theory.

[21] A serious omission in Lewis PW, for example is any definition or characteristic of *world* ('world' does not even appear as an index item). At most we are offered, no adequate substitute for an account, some heavy and hardly evident restrictions on worlds and what they can contain. It is in part because of lack of characterisation of worlds that Lewis is open to such objections as that constructions from his plurality of worlds afford further worlds incompatible with other features of his theory, objections he makes heavy weather of meeting (despite the "no-hands" appearance of his responses: e.g. p. 107ff.).

It is in part because of this lack of account too that he can suppose, when it suits, that 'world' behaves like 'mountain' — a comparison through which he tries to exclude impossible worlds (p. 7). Connectives "distribute through" the modifier 'on the mountain', so that the (odd) "on the mountain, *p* and not-*p*" yields the outright contradiction, *p* (on the mountain) and not-*p* (on the mountain). Similarly, so Lewis assumes, for worlds, impossible worlds would yield flat-out contradiction. But the comparison is badly flawed, even on the information Lewis does supply. It is not just that mountains would suffer serious incompleteness as worlds, but worse that the same people can be at or on many different mountains though not (according to Lewis) in several different worlds. More important, various functors distribute through 'on the mountain' which do *not* pass through 'in a world' on a theory such as Lewis is propounding, notably 'if' (consider, e.g., 'on the mountain if it rains a parka could be advisable'), 'possibly', etc. Lewis's objection if pressed would collapse possible worlds theory towards a one-world position.

[22] The answer meets a problem posed in Nola, p. 331. According to Nola, rejecting the assumption of a unique correct world makes one a radical idealist! (p. 321). But why accept this realist/idealist distinction? It depends on the mistaken ontological assumption that terms that do not refer are not about anything.

[23] Once again, the whole ontologizing business is criticised, in detail, in JB, where noneism is elaborated. The replacement of ontology by sistology is explained in SF.

[24] Historically realism has come to involve a complex mesh of interrelated contrasts. As should be evident, the concern here is not with realism in perception, nor except incidentally with realism concerning universals. It is with realism concerning things, and worlds especially.

[25] I am much indebted to Robert Nola and to an anonymous referee for detailed comments on an earlier version of this article. I have also profited much from discussions with Len Goddard, Graham Priest, and David Bennett, though we have subsequently gone our different worldly ways.

REFERENCES

Barrow, J. D. and Tipler, F. J. (1986) *The Anthropic Cosmological Principle*, Oxford: Clarendon Press.

Berlin, I. (1976) *Vico and Herder*, London: Hogarth Press.

Bloor, D. (1976) *Knowledge and Social Imagery*, London: Routledge & Kegan Paul.

Bohm, D. (1980) *Wholeness and the Implicate Order*, London: Routledge & Kegan Paul.

Blackburn, S. (1984) *Spreading the Word*, Oxford: Clarendon.

Davies, P. (1983) *God and the New Physics*, London: Dent.

Descartes, R. *A Discourse on Method*.

Devitt, M. (1984) *Realism and Truth*, Princeton: Princeton University Press.

Edwards, J. (1982) *Ethics without Philosophy*, Tampa: University Press of Florida.

Fine, A. (1984) 'And Not Anti-Realism Either', *Nous* **18**, 51—65.

Goodman, N. (1978) *Ways of Worldmaking*, Brighton: Harvester Press.

Hooker, C. A. (1986) *A Realistic Theory of Science*, New York, State University of New York Press.

Joseph, G. (1977) 'Conventionalism and Physical Holism', *Journal of Philosophy* **74**, 439—62.

Knorr-Cetina, K. D. and Malkay, M. (eds.) (1983) *Science Observed*, London: Sage.

Laudan, L. (1984) 'A Confutation of Convergent Realism'; reprinted in Lepin (1984).

Leplin, J. (ed.) (1984) *Scientific Realism*, Berkeley: University of California Press.

Lewis, D. (1985) *On the Plurality of Worlds*, Oxford: Blackwell; referred to as PW.

Moore, G. E. (1959) 'Proof of an External World', *Philosophical Papers*, London: Allen & Unwin.

Nola, R. (1980) 'Paradigms Lost, or The World Regained —', *Synthese* **45**, 317—50.

Oxford English Dictionary, Oxford, Oxford University Press, 1971; referred to as OED.

Plumwood, V. and Routley, R. (1982) 'The Inadequacy of the Actual and the Real: Beyond Empiricism, Idealism and Mysticism', in *Language and Ontology* (ed. W. Leinfellner and others), Vienna, Hölder-Pichler-Tempsky: 49—87; referred to as WS.

Putnam, H. (1980) 'Models and Reality', *Journal of Symbolic Logic* **45**, 464—82.

Putnam, H. (1981) *Reason, Truth and History*, Cambridge: Cambridge University Press; referred to as RH.

Rescher, N. (1978) 'Philosophical Disagreement: An Essay Towards Orientational Pluralism in Metaphilosophy', *Review of Metaphysics* **32**, 217—51.

Rescher, N. (1985) *The Strife of Systems*, Pittsburgh: University of Pittsburgh Press.

Rorty, R. (1972) 'The World Well Lost', *Journal of Philosophy* **89**, 649—669.

Routley, R. (1976) 'The Semantical Metamorphosis of Metaphysics', *Australasian Journal of Philosophy* **54**, 187—205; referred to as SM.

Routley, R. (1979) *Exploring Meinong's Jungle and Beyond*, Canberra: RSSS, Australian National University; referred to as JB.

Routley, R. (1981) 'Necessary Limits to Knowledge: Unknowable Truths', in *Essays in Scientific Philosophy* (ed. E. Morscher and others), Bad Reichenhall: Comes Verlag: 93—115: referred to as NL.

Segerberg, K. (1971) *An Essay in Classical Modal Logic*, Uppsala: Stockholm Studies in Philosophy.

Shapin, S. (1982) 'History of Science and its Sociological Reconstruction', *History of Science* **20**, 157–211.

Stove, D. (0000) 'Epistemology and the Ishmael Effect', typescript, University of Sydney; to appear.

Sylvan, R. (1984) 'Philosophy, Politics and Pluralism I. Relevant Modellings and Arguments', *Research Series in Logic and Metaphysics*, Canberra: RSSS, Australian National University; also in *Journal of Non Classical Logic* **4** (1987), 57–107; referred to as PP.

Sylvan, R. (1984) 'How Science and Fiction and Myth Step Beyond the Actual', *Research Series in Unfashionable Philosophy*, Canberra: RSSS, Australian National University; also in *Zeitschrift für Semiotik* **9** (1987), 129–52; referred to as SF.

Sylvan, R. (1985) 'Science and Science: Relocating Stove and the Modern Irrationalists' *Critical Philosophy* **2**, 16–28; referred to as SS.

Sylvan, R. (1985–6) 'Towards and Improved Cosmological Synthesis', *Grazer Philosophische Studien* **25/26**, 135–79; referred to as CS.

Sylvan, R. (1987) 'Establishing the Correspondence Theory of Truth and Rendering it Coherent', to appear in *Stephan Körner: Contributions to Philosophy*, referred to as SC.

Sylvan, R. (0000). 'On Making a Coherence Theory of Truth True', typescript, Canberra; referred to as CT.

Waismann, F. (1945–6) 'Are There Alternative Logics?', *Proceedings of the Aristotelian Society* n.s. **46**; 77–104.

Wisdom, J. (1952) *Other Minds*, Oxford: Blackwell.

Wittgenstein, L. (1969) *On Certainty*, Oxford: Blackwell.

Woolgar, S. (1983) 'Irony in the Social Study of Science', in Knorr-Cetina, K. D. and Mulkay, M. (eds.) (1983).

Ziman, J. (1978) *Reliable Knowledge*, Cambridge: Cambridge University Press.

NOTES ON CONTRIBUTORS

GREGORY CURRIE was born in England and educated at the London School of Economics. He teaches at the University of Otago, New Zealand. His publications include *Frege: An Introduction to his Philosophy* (1982) and *An Ontology of Art Works* (forthcoming). He has published a number of papers on Frege's philosophy, the philosophy of science, the philosophy of language and aesthetics.

JOHN FOX was born in Wellington in 1941. From 1958 through 1966 he was a Jesuit. He studied the history and philosophy of science at the University of Melbourne, and since 1967 has taught philosophy at La Trobe University. His interests include the philosophy of science, the philosophy of logic and peace.

C. A. HOOKER studied physics (Ph.D. Sydney University, 1968) and then philosophy (Ph.D. York University, Toronto, 1970), taught philosophy and engineering decision making in Canada for a decade and has been Professor of Philosophy at the University of Newcastle, Newcastle, Australia since 1980. He is the co-author of *Energy and the Quality of Life: Understanding Energy Policy* (1980), *Images of Science* (1986), the author of *A Realistic Theory of Science* (1987) and the editor of a dozen books concerned with philosophy of science and/or science policy. Needless to add, numerous articles published and honourable institutional services performed have competed for his time with being a husband, father, friend and nature lover.

KAI HAHLWEG was educated in Germany and Canada and is at present post-doctoral fellow at the University of Newcastle, Australia. He studied chemistry in Munich and Philosophy at the University of Western Ontario, concentrating on the History and Philosophy of Science, in particular philosophy of biology and history of chemistry. In his Ph.D. thesis, 'The Evolution of Science: A Systems Approach' (Western Ontario 1983) he developed a new evolutionary epistemology which is outlined in his 'Progress and Stability', forthcoming in *Biology and Philosophy*.

Robert Nola (ed.), Relativism and Realism in Science, 293–295.

FRED KROON was born in the Netherlands in 1947, and educated at the University of Auckland and Princeton University. Since 1974 he has been teaching philosophy at the University of Auckland, where he is currently a Senior Lecturer. His publications have mostly been in the area of epistemic and semantic paradox and the theory of reference.

LARRY LAUDAN occupies the chair of Philosophy at the University of Hawaii. He has taught in English, German, Australian and American universities. His publications include *Progress and Its Problems, Science and Hypothesis*, and *Science and Values.*

ALAN MUSGRAVE was born in England. He studied and then taught at the London School of Economics from 1958 until 1970, receiving his doctoral degree in philosophy in 1969. Since 1970 he has been Professor of Philosophy at the University of Otago in New Zealand. He and Imre Lakatos edited *Problems in the Philosophy of Science* (1968) and *Criticism and the Growth of Knowledge* (1970); he and Greg Currie edited *Popper and the Human Sciences* (1985). He has published numerous papers in the philosophy and history of science, most of the recent ones devoted to the defence of a realist view of science.

GRAHAM ODDIE was born in New Zealand in 1954, and studied at the University of Otago and the London School of Economics. He has taught at the University of Otago and is currently Professor at Massey University. He has written articles on metaphysics, philosophy of science, and philosophical logic, and is the author of *Likeness to Truth* (1986), the first book-length work on truthlikeness.

DAVID PAPINEAU was born in 1947 and educated in Trinidad, England and South Africa. After reading mathematics at the University of Natal and philosophy at King's College, Cambridge, he did a Ph.D. in the philosophy of science at Cambridge. He has held posts at Reading University, at Macquarie University, Sydney, and at Birkbeck College, University of London. He is currently a lecturer in the Department of History and Philosophy of Science, Cambridge University. His publications include *For Science in the Social Sciences* (1978), *Theory and Meaning*, (1979), and *Reality and Representation* (1987).

PHILIP PETTIT was born in Ireland in 1945 and studied philosophy at the National University of Ireland and for his Ph.D. at Queen's University, Belfast. He lectured at University College Dublin and held a Research Fellowship at Trinity Hall, Cambridge, before being appointed in 1977 as Professor of Philosophy at Bradford University. In 1983 he moved to his present position as Professorial Fellow at the Research School of Social Sciences, Australian National University, Canberra. His publications include: *The Concept of Structuralism* (1975), *Judging Justice* (1980) and with Graham Macdonald, *Semantics and Social Science* (1981).

RICHARD SYLVAN an Antipodean researcher presently resident in Australia, is trying to put together a systematic and comprehensive *deep theory*, including among other parts, deep green theory and its practice, deep pluralism and its regional applications, deep relevant logic and its dialectical elaborations, all set within integrated object and process theory. But he also contributes, when not in the forest or field, to the endless negative, often exhausting business of criticism — intellectual, social and environmental. In particular, he is a persistent untiring critic of prevailing dogmatic and reductionist philosophies, such as materialism, empiricism, extensionalism, utilitarianism, economism and, not least, Australian primitivism. Richard is a director of the Ecological Foundation, and can be reached at Stonywood Vinyard, Forest Road, Bungendore, Australia 2621.

The Editor

ROBERT NOLA was born in New Zealand and studied at the University of Auckland and The Australian National University. He currently teaches at the University of Auckland. His publications have been mainly in the philosophy of science and on historical figures including Plato, Marx and Nietzsche.

INDEX OF NAMES

AUSTRALASIAN STUDIES IN HISTORY AND PHILOSOPHY OF SCIENCE

General Editor:

R. W. HOME, *University of Melbourne*

1. Robert McLaughlin (ed.), *What? Where? When? Why? Essays on Induction, Space and Time, Explanation.* 1982, xvii + 319 pp.
2. David Oldroyd and Ian Langham (eds.), *The Wider Domain of Evolutionary Thought.* 1983, xx + 326 pp.
3. R. W. Home (ed.), *Science Under Scrutiny: The Place of History and Philosophy of Science.* 1983, xvii + 182 pp.
4. John A. Schuster and Richard R. Yeo (eds.), *The Politics and Rhetoric of Scientific Method. Historical Studies.* 1986, xxxix + 305 pp.
5. John Forge (ed.), *Measurement, Realism and Objectivity: Essays on Measurement in the Social and Physical Sciences.* 1987, xvii + 346 pp.